NUMERICAL COMPUTATION OF ELECTRIC AND MAGNETIC FIELDS

NUMERICAL COMPUTATION OF ELECTRIC AND MAGNETIC FIELDS

Charles W. Steele
Consultant

VNR VAN NOSTRAND REINHOLD COMPANY
New York

Library of Congress Catalog Card Number 85-22659
ISBN 0-442-27841-1

Printed in the United States of America.

Van Nostrand Reinhold Company Inc.
115 Fifth Avenue
New York, New York 10003

Van Nostrand Reinhold Company Limited
Molly Millars Lane
Wokingham, Berkshire RG11 2PY, England

Van Nostrand Reinhold
480 La Trobe Street
Melbourne, Victoria 3000, Australia

Macmillan of Canada
Division of Canada Publishing Corporation
164 Commander Boulevard
Agincourt, Ontario M1S 3C7, Canada

16 15 14 13 12 11 10 9 8 7 6 5 4 3 2

Library of Congress Cataloging-in-Publication Data
Steele, Charles W.
Numerical computation of electric and magnetic fields.
Includes bibliographies and index.
1. Electric fields—Measurement. 2. Magnetic fields—Measurement. 3. Numerical analysis. I. Title.
QC665.E38S74 1986 537'.01'511 85-22659
ISBN 0-442-27841-1

To
CANDACE

PREFACE

For well over a decade, the numerical approach to field computation has been gaining progressively greater importance. Analytical methods of field computation are, at best, unable to accommodate the very wide variety of configurations in which fields must be computed. On the other hand, numerical methods can accommodate many practical configurations that analytical methods cannot. With the advent of high-speed digital computers, numerical field computations have finally become practical.

However, in order to implement numerical methods of field computation, we need algorithms, numerical methods, and mathematical tools that are largely quite different from those that have been traditionally used with analytical methods. Many of these algorithms have, in fact, been presented in the large number of papers that have been published on this subject in the last two decades. And to some of those who are already experienced in the art of numerical field computations, these papers, in addition to their own original work, are enough to give them the knowledge that they need to perform practical numerical field computations.

But newcomers to the art of numerical field computations need a more orderly presentation of the necessary background information than is available to them in these papers. The objective of this book is precisely to provide this orderly presentation of the necessary information. Clarity of presentation is given greater emphasis than a high degree of sophistication or being up to the state of the art. It is my hope that those who read this book will then be able to read and understand the recent papers and thereby achieve the level of sophistication to which they aspire.

Specifically, this book has the objective of presenting information in two different categories:

1. Certain algorithms that have been used successfully for computation of fields.
2. Background information that will help readers to develop their own algorithms.

Putting it another way, the intent of this book is to help readers develop or select the algorithms to be used in these field computations. These algorithms can then be used in computer programs, and these computer programs can be run in high-speed digital computers to compute the fields.

To build these computer programs, the programmer requires knowledge in a number of other areas beyond the algorithms that are discussed in this book. These areas include computer programming, methods for numerical solution of systems of equations, methods for automatic deployment of node points, and construction of finite elements and computer graphics (to display the computed fields). These important topics are discussed in technical journals as well as in other books.

Any numerical field computation is built upon logic that is necessarily an intimate mixture of mathematics, numerical analysis, and electromagnetic theory. This book is just such a mixture. It is assumed that the reader has some prior knowledge of electromagnetic theory, vector analysis, and linear algebra.

This book covers static and quasi-static field problems in media that generally vary in permittivity, permeability, and conductivity. It does not cover dynamic problems, such as antennas, scatterers, radiation in free space, and waveguide modes. These items are topics for books in themselves.

Here we consider only problems in which both the observer of the fields and the medium throughout the entire domain are fixed with respect to a single frame of reference. That is, we do not consider problems in which a portion of the medium moves with respect to the remainder of the medium (and the field is computed in both)—for example, problems in which the field is to be computed in both the moving rotor and stator of an electric motor or generator. Such problems are an important study in themselves.

An objective of this book has been to use standard nomenclature as much as possible. All of the equations related to electric and magnetic fields are based upon the use of the SI system of units. The symbols used are, for the most part, taken from the international recommendations for quantities, units, and their symbols, as published by the United States of America Standards Institute.

I am deeply indebted to a number of people for their contributions to this book. I appreciate my discussion with Mr. Charles Trowbridge, of Rutherford Laboratories, United Kingdom, regarding my approach to Chapter 5. Dr. Alvin Wexler of the University of Manitoba made valuable suggestions related to my presentation of certain of his material in Chapter 7. In addition, Mr. Trowbridge and Dr. Wexler provided thorough technical reviews of my completed manuscript. Dr. Dennis Lindholm of Ampex Corporation reviewed the majority of the manuscript and made many valuable suggestions. Mrs. Bernie Jones of Palo Alto, California, did an excellent job of typing a large portion of the manuscript.

Finally, and most important, I wish to acknowledge my wife, Candace. She has gently tolerated my work on this book over four annual vacations and countless weekends. Without her unfailing support and encouragement I could not have written it.

If the reader wishes to communicate with the author regarding this book, he can send the communication to the publisher, who will then forward it to the author.

CHARLES W. STEELE

CONTENTS

Preface / vii

1. Introduction / 1

2. Field Properties / 3

2.1 Introduction / 3

2.2 Maxwell's Equations in the Dynamic, Quasi-Static, and Static Cases / 3

2.3 Polarization and Magnetization / 6

2.4 Laws for Static Fields in Unbounded Regions / 9

2.5 Integral Representations for Quasi-Static Fields Using the Helmholtz Theorem / 12

2.6 Equivalent Configurations / 20

2.7 Steady-State Dynamic Problems and Phasor Field Representations / 21

2.8 Continuity Conditions of Fields at a Medium Discontinuity / 23

References / 27

3. Problem Definition / 28

3.1 Introduction / 28

3.2 Field Problem Domains, Source Problem Domains, Interior Problems, and Exterior Problems / 29

3.3 Is the Problem Static, Quasi-Static, or Dynamic? / 30

3.4 What Field Is To Be Computed? / 31

3.5 Is the Problem Two-Dimensional or Three-Dimensional? / 31

3.6 The Medium / 31

3.7 Boundary Conditions / 33

4. Linear Spaces in Field Computations / 35

4.1 Introduction / 35

4.2 Basis Functions / 36

4.3 Shape Functions / 41

4.4 Finite Elements and Shape Functions of Global Coordinates in Two-Dimensional Problem Domains / 42

4.5 Isoparametric Shape Functions in Two Dimensions / 51
4.6 Finite Elements and Shape Functions of Global Coordinates in Three-Dimensional Problem Domains / 55
References / 58

5. Projection Methods in Field Computations / 59
5.1 Introduction / 59
5.2 Special Linear Spaces in Field Computations / 61
5.3 Operators in Field Calculations / 64
5.4 Approaches Used in Obtaining Approximate Solutions to Field Problems / 65
5.5 Finite Element Method for Interior Problems / 73
5.6 Integral Equation Method / 77
5.7 Projection Methods / 80
5.8 Orthogonal Projection Methods / 85
References / 90

6. Finite Element Method for Interior Problems / 91
6.1 Introduction / 91
6.2 Formulation of Finite Element Method for Interior Problems / 92
6.3 Computation of Linear System for Finite Element Method / 98
6.4 Sample Problem / 105
References / 110

7. Finite Element Method for Exterior Problems / 111
7.1 Introduction / 111
7.2 McDonald-Wexler Algorithm / 111
7.3 Silvester et al. Algorithm / 120
References / 133

8. Integral Equation Method / 134
8.1 Introduction / 134
8.2 Linear and Uniform Media in Continuity Subdomains / 135
8.3 Saturable, Nonlinear, and Nonuniform Media in Continuity Subdomains / 142
8.4 Numerical Solution of Integral Equations – General Approach / 143
8.5 Finite Elements and Basis Functions Used in the Integral Equation Method / 148

8.6 Integral Equation Numerical Solution by the Collocation Method / 149
8.7 Integral Equation Numerical Solution by the Galerkin Method / 151
8.8 Numerical Integration / 156
8.9 Sample Problem / 159
References / 165

9. Static Magnetic Problem / 166
9.1 Introduction / 166
9.2 Interior Static Field Problems / 167
9.3 Exterior Static Problems Approximated by Interior Problems / 167
9.4 Exterior Magnetic Static Problem / 170
9.5 Static Magnetic Field in a Saturable Medium / 173
References / 174

10. Eddy Current Problem / 175
10.1 Introduction / 175
10.2 Commonly Used Basic Formulations for the Eddy Current Problem / 176
10.3 Simple Two-Dimensional Eddy Current Problem / 181
10.4 Projection Methods for General Eddy Current Problems / 182
10.5 Eddy Current Problem Formulation / 189
References / 196

Glossary / 197

Appendix A Derivation of the Helmholtz Theorem / 201
Appendix B Properties of the Magnetic Vector Potential, **A** / 211
Appendix C Integral Expressions for Scalar Potential from Green's Theorem / 213

Index / 221

NUMERICAL COMPUTATION OF ELECTRIC AND MAGNETIC FIELDS

1
INTRODUCTION

For the last 20 years, the high-speed digital computer has been revolutionizing the computation of electric and magnetic fields, to the point that most practical computations of fields are now done numerically on a computer. This is because many, if not most, practical problems that arise in engineering and science can be solved numerically but cannot be solved analytically. And the computer is practically necessary for numerical solutions.

Between 1873, when Maxwell announced his field equations, and, say, 1950, an enormous amount of excellent work went into obtaining analytical solutions to field problems. A wide variety of coordinate systems were developed and both static and dynamic field solutions were developed in these coordinate systems. Since these solutions have the property of orthogonality, it is possible, for a given problem, to synthesize a linear combination of such solutions to meet a specified boundary condition. Even so, the variety of coordinate systems and of boundary conditions thereby afforded is not adequate to deal with many practical problems. Thus, many practical problems that cannot be solved by analytical methods can be solved by numerical methods.

As a result, a new science is emerging—the science of numerically computing electric and magnetic fields. This science is necessarily a mixture of electromagnetic theory, mathematics, and numerical analysis. In fact, each *problem* of numerical computation of fields is an intimate mixture of these three disciplines.

The objective of this book is to present information that the reader needs to solve certain field problems numerically. This presentation is then a mixture of electromagnetic theory, numerical analysis, and mathematics.

This book is limited to static and quasi-static field problems.

Any field computation necessarily starts with consideration of the properties of the fields themselves. Accordingly, Chapter 2 reviews electromagnetic theory to the extent necessary to provide the basis in physics of the field problems considered in this book.

Chapter 3 provides a unified treatment for the definition of field problems. This chapter is intended to underscore the importance of having a clear definition of the problem at the outset.

In most of our numerical field computations, our solution can best be thought of as an element of a finite-dimensional linear space that we construct ourselves. Chapter 4 discusses, in some detail, the construction of these linear spaces.

Since most of the algorithms that we use in field computations are projection methods, Chapter 5 provides a unified treatment for the different projection methods that we use.

Chapters 2, 3, 4, and 5 provide support for the material presented in the subsequent chapters.

Chapters 6, 7, and 8 discuss the two most commonly used methods of numerical field computation, the finite element method and the integral equation method.

Chapters 9 and 10 present and discuss specific algorithms that have been used in the numerical solution of static and quasi-static field problems.

Readers of a book on this topic can have different interests in field computation. First, there are those for whom the interest in field computations is secondary to a primary objective, such as experimental research or the development of a device. These readers want to make good, practical field computations. They do not feel justified in spending time and effort in inquiring into the basic nature of field computations.

A second group of people are primarily interested in the numerical computation of fields *itself*. These are people who are interested in advancing the state of the art in field computations.

This second group has more interest in understanding the basic theory underlying field numerical field computations than does the first group. An objective of this book is to serve both groups. Chapter 5 is intended primarily for the second group. The author hopes that those in the first group can get what they need from this book without a careful study of this chapter.

Another objective of this book is to make it unnecessary for any reader to read the whole book (unless, of course, he or she *wants* to). It is for this reason that a glossary of the most important and commonly used symbols is included. Furthermore, at points where latter chapters require support from former chapters, the appropriate references to the former material are made.

2
FIELD PROPERTIES

2.1. INTRODUCTION

This chapter develops and presents the equations in electromagnetic theory that provide the basis for the computational methods and algorithms presented in later chapters.

Nonuniformities and discontinuities in the media (the permeability, permittivity, and conductivity) are a major problem in practical field computations. This is true whether the computations are made analytically or numerically. For numerical computations, these nonuniformities can tax the skill, patience, and endurance of the scientist, and the capacities of his computer as well (in terms of high-speed memory capacity and computational speed and time). To cope with this problem, a variety of methods for dealing with nonuniformities have been presented in the published papers on numerical field computations.

This chapter takes the approach of the aggregate of these papers. To deal with the problem of these nonuniformities, the chapter presents a variety of approaches and theoretical formulations. Emphasis is placed on developing formulations that form the basis for practical numerical algorithms. Emphasis is also placed on developing ways for the scientist to develop a physical "feel" for the behavior of the fields, since a good feel for field behavior is an essential ingredient for a successful field computation.

Emphasis is placed on a careful presentation of the differences between static, quasi-static, and dynamic field behaviors. These differences are extremely important in practical numerical field computations. Static field computations are simpler than quasi-static field computations, and quasi-static field computations are simpler than dynamic field computations. And for field computations, it is *always* best to use the simplest acceptable approach.

2.2. MAXWELL'S EQUATIONS IN THE DYNAMIC, QUASI-STATIC, AND STATIC CASES

2.2.1. Dynamic Case

The methods or algorithms discussed in this book are based largely upon Maxwell's equations. These equations, for the fully dynamic case, are the following:

$$\nabla \times \mathbf{H} = \mathbf{J} + \frac{\partial \mathbf{D}}{\partial t} \tag{2.2.1-1}$$

$$\nabla \times \mathbf{E} = -\frac{\partial \mathbf{B}}{\partial t} \tag{2.2.1-2}$$

$$\nabla \cdot \mathbf{B} = 0 \tag{2.2.1-3}$$

$$\nabla \cdot \mathbf{D} = \rho \tag{2.2.1-4}$$

In these equations **H** and **E** are the magnetic and electric fields, **J** is the conduction current density and ρ is the electric charge density. Furthermore, **B** is the magnetic flux density and **D** is the displacement, or electric flux, density.

Integral forms can derived from the four Maxwell's equations. Using Stokes' Theorem, we derive from Equations (2.2.1-1) and (2.2.1-2) that

$$\oint_C \mathbf{H} \cdot \mathbf{dl} = I + \frac{\partial \Psi}{\partial t} \tag{2.2.1-5}$$

$$\oint_C \mathbf{E} \cdot \mathbf{dl} = -\frac{\partial \Phi}{\partial t} \tag{2.2.1-6}$$

The integrals in these equations are line integrals around the closed contour C. In other words, they give the *circulation* of **H** and **E** around the path C. The symbols Ψ and Φ represent the *electric flux* and the *magnetic flux* that thread path C, and I is the conduction current that flows through path C. Using the divergence theorem, we derive from Equations (2.2.1-3) and (2.2.1-4), that

$$\oint_S \mathbf{B} \cdot \mathbf{dS} = 0 \tag{2.2.1-7}$$

$$\oint_S \mathbf{D} \cdot \mathbf{dS} = Q \tag{2.2.1-8}$$

where Q is the electric charge inside the surface S. The integrals in Equations (2.2.1-7) and (2.2.1-8) give the *outflow* of vectors **B** and **D** over the closed surface S. The symbol Q is the free electric charge enclosed within the closed surface S.

These equations, in this fully dynamic form, are needed for computations of radiated fields, that is, fields radiated and received by antennas, fields radiated down waveguides, and fields in electronic devices such as klystrons and magnetrons. For many other applications, simpler approximate versions of these equations are adequate.

2.2.2. Quasi-Static Case

The quasi-static case differs from the fully dynamic case only by its neglect of the displacement current.* That is, in the quasi-static case, we say that

$$\nabla \times \mathbf{H} = \mathbf{J} \tag{2.2.2-1}$$

and that

$$\oint_C \mathbf{H} \cdot \mathbf{dl} = I \tag{2.2.2-2}$$

and that Equations (2.2.1-2), (2.2.1-3), (2.2.1-4), (2.2.1-6), (2.2.1-7), and (2.2.1-8) still hold. Since the divergence of the curl of a vector is zero, we see from (2.2.2-1) that in the quasi-static case,

$$\nabla \cdot \mathbf{J} = 0 \tag{2.2.2-3}$$

The quasi-static approximation is used for time-varying fields in many conducting media. This is because, for good conductors, the conduction current greatly exceeds the displacement current for frequencies that usually concern us (right up to X-ray frequencies). A good example of this is the calculation of time-varying magnetic fields in iron cores, the so-called "eddy-current" problem. This has practical applications in electric motors, generators, magnetic recording heads, and solenoid actuators.

2.2.3. Static Case

In the static case, both the electric displacement current *and* the time-varying magnetic flux density are neglected. In this case, then, Equation (2.2.1-2)

---

* The displacement current can be neglected when

$$\sigma \gg \omega e$$

where σ, ω, and e are the conductivity, radian frequency, and permittivity.

becomes

$$\nabla \times \mathbf{E} = 0 \tag{2.2.3-1}$$

and Equation (2.2.1-6) becomes

$$\oint_C \mathbf{E} \cdot \mathbf{dl} = 0 \tag{2.2.3-2}$$

In addition to these equations, Equations (2.2.1-3), (2.2.1-4), (2.2.1-7), (2.2.1-8), (2.2.2-1), and (2.2.2-2) still hold.

Static field calculations are made wherever the dictates of physical reality permit, because, as we will see below, they are of relative simplicity compared to dynamic and quasi-static calculations. As an example, many calculations in the magnetic fields of magnets are static calculations.

2.3. POLARIZATION AND MAGNETIZATION

2.3.1. Polarization

Figure 2-1 shows an electric dipole which consists of two point charges, $-Q$ and $+Q$ (equal in magnitude and opposite in sign), that are separated by a distance d. There is the vector $\mathbf{d}$, of magnitude d, and pointing in the direction toward the positive charge and away from the negative charge. This dipole has an *electric dipole moment* $\mathbf{p}$, given by

$$\mathbf{p} = Q\mathbf{d} \tag{2.3.1-1}$$

Suppose that there is a system of charges (say, a molecule) that is neutral (i.e., has a total charge of zero). Since the sum of all positive charges in the system equals in magnitude the sum of all negative charges, then that molecule

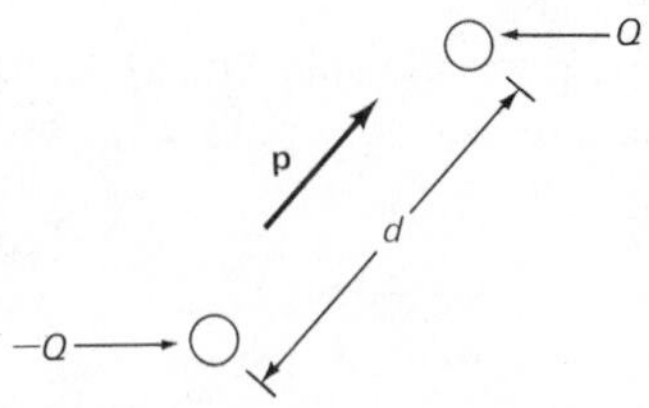

Fig. 2-1. Electric dipole.

can be considered to consist of a number of elementary electric dipoles, as described above. This molecule has an electric dipole moment equal to the vector sum of the moments of these elementary electric dipoles. Suppose that within a small volume, V, there are M molecules, each with a dipole moment, $\mathbf{p}_m$. Then we define the *polarization density*, or *polarization*, as given by

$$\mathbf{P} = \frac{1}{V} \sum_{m=1}^{M} \mathbf{p}_m \tag{2.3.1-2}$$

The practical value of this definition of polarization lies in the fact that, within a small region having a uniform dielectric medium, the *value* of $\mathbf{P}$ is very nearly independent of which small volume V is chosen. Thus, $\mathbf{P}$, as defined, has meaning on a *macroscopic* scale. Notice that $\mathbf{P}$ can be thought of as being the result of a number of infinitesimal molecular dipole moments *acting in a vacuum.*

The displacement density, D, in any material is given by

$$\mathbf{D} = \varepsilon_0 \mathbf{E} + \mathbf{P} \tag{2.3.1-3}$$

where ε_0 is the permittivity of free space. If the material is linear, then

$$\mathbf{P} = \underline{\bar{\chi}}_e \varepsilon_0 \mathbf{E} \tag{2.3.1-4}$$

where $\underline{\bar{\chi}}_e$ is dimensionless and is the electric susceptibility tensor. If the material is linear *and* isotropic,

$$\mathbf{P} = \chi_e \varepsilon_0 \mathbf{E} \tag{2.3.1-5}$$

where χ_e is the susceptibility, a dimensionless scalar. From Equations (2.3.1-3) and (2.3.1-4) for linear materials,

$$\mathbf{D} = \varepsilon_0 (1 + \underline{\bar{\chi}}_e) \mathbf{E} \tag{2.3.1-6}$$

From Equations (2.3.1-3) and (2.3.1-5), for a linear *and* isotropic medium,

$$\mathbf{D} = \varepsilon_0 (1 + \chi_e) \mathbf{E} = \varepsilon_0 \varepsilon_r \mathbf{E} = \varepsilon \mathbf{E} \tag{2.3.1-7}$$

where

$$\varepsilon = \varepsilon_0 (1 + \chi_e)$$

is the scalar permittivity, and ε_r is the *relative* permittivity.

2.3.2. Magnetization

The magnetization, M, is defined in much the same way that the polarization was. A molecule (or an atom) has a magnetic dipole moment, m, which results from its electrons in orbit and its spinning electrons.

This magnetic dipole moment, m, could alternatively have been produced by a certain current, I, flowing in a certain circular loop, as shown in Figure 2-2. Therefore, the total dipole moment that results from any number of molecules could, equivalently, be produced by a number of equivalent current loops of the type shown in Figure 2-2, *operating in a vacuum.* The magnetic dipole moment, m, could also have been produced by two fictitious magnetic point charges, of equal magnitudes and opposite polarities, separated by some distance, similar to the electric dipole shown in Figure 2-1. Therefore, the total dipole moment that results from any number of molecules could, equivalently, be produced by a magnetic charge density acting *in a vacuum.*

If there is a collection of N molecules within a small volume, V, then, on a *macroscopic* scale, there is a magnetization density, or magnetization, given by

$$\mathbf{M} = \frac{1}{V}\sum_{i=1}^{N} \mathbf{m}_i \qquad (2.3.2\text{-}1)$$

where $\mathbf{m}_i$ is the magnetic dipole moment the ith molecule in the volume V.

The magnetization flux density, $\mathbf{B}$, is given by

$$\mathbf{B} = \mu_0(\mathbf{H} + \mathbf{M}) \qquad (2.3.2\text{-}2)$$

If the material is linear, then

$$\mathbf{M} = \underline{\chi}\mathbf{H} \qquad (2.3.2\text{-}3)$$

where $\underline{\chi}$ is dimensionless and is the magnetic susceptibility tensor. If the material is linear *and isotropic,*

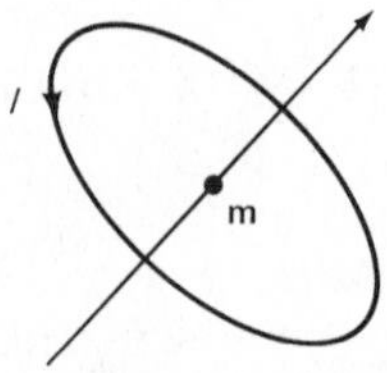

Fig. 2-2. Magnetic dipole.

$$\mathbf{M} = \chi\mathbf{H} \tag{2.3.2-4}$$

where χ is dimensionless and is the scalar magnetic susceptibility. From Equations (2.3.2-2) and (2.3.2-3), for linear material,

$$\mathbf{B} = \mu_0(1 + \underline{\chi})\mathbf{H} \tag{2.2.2-5}$$

From Equations (2.3.2-2) and (2.3.2-4) for linear *and* isotropic media,

$$\mathbf{B} = \mu_0(1 + \chi)\mathbf{H} = \mu_0\mu_r\mathbf{H} = \mu\mathbf{H} \tag{2.3.2-6}$$

where μ is the permeability, and μ_r is the relative permeability.

2.4. LAWS FOR STATIC FIELDS IN UNBOUNDED REGIONS

2.4.1. Coulomb's Law

Coulomb's law states that the force, F, between two point charges, Q_1 and Q_2, is

$$F = \frac{Q_1 Q_2}{4\pi\varepsilon_0 r^2} \tag{2.4.1-1}$$

where r is the distance between the two charges. If we define the electrostatic field (the electric field in the static case) as the force applied by one charge on a unit positive point charge, we have from Equation (2.4.1-1) that

$$\mathbf{E} = \frac{Q\mathbf{a}_r}{4\pi\varepsilon_0 r^2} \tag{2.4.1-2}$$

The point at which the charge Q is located is called the *source point*, and the point at which the field $\mathbf{E}$ is taken is called the *field point*. These terms are to find frequent usage throughout the remainder of this book. Then r is the distance between the source point and the field point, and $\mathbf{a}_r$ is a unit vector in the direction *from* the source point *toward* the field point.

Now we define the electrostatic potential, ϕ_e, as the energy required to bring a unit positive charge from infinity to the field point. Then ϕ_e at point P is given by

$$\phi_e = \int_{\infty}^{P} \mathbf{F}\cdot d\mathbf{l} = \int_{\infty}^{P} \mathbf{E}\cdot d\mathbf{l} \tag{2.4.1-3}$$

Since, from (2.4.1-3) we have that

$$\phi_e = -\int_P^\infty \mathbf{E}\cdot d\mathbf{l} \tag{2.4.1-4}$$

then from this equation

$$\mathbf{E} = -\nabla\phi_e \tag{2.4.1-5}$$

Substituting Equation (2.4.1-2) into Equation (2.4.1-4), we have the electrostatic potential field about a single point charge, Q, given by

$$\phi_e = \frac{Q}{\varepsilon_0 4\pi r} \tag{2.4.1-6}$$

The electrostatic field and electrostatic potential resulting from any number of point charges in a free-space environment can be obtained by adding up the single-charge fields as given by Equations (2.4.1-2) and (2.4.1-6). By using integrals to represent the limits of such summations, we can derive from Equations (2.4.1-2) and (2.4.1-6) the corresponding equations for the fields from an electric charge density. Consider the configuration in Figure 2-3. In this figure, we have a charge cloud of charge density ρ enclosed within the sphere S. Outside S, there is no charge, *only* free space. This universe is assumed to contain no dielectrics, only the charge density and perhaps conductors in an otherwise free-space environment. Starting with Equations (2.4.1-2) and (2.4.1-6), we see that the electrostatic field, $\mathbf{E}$, and the electrostatic potential, ϕ_e, at any point P (either inside or outside S) in Figure 2-3 are given

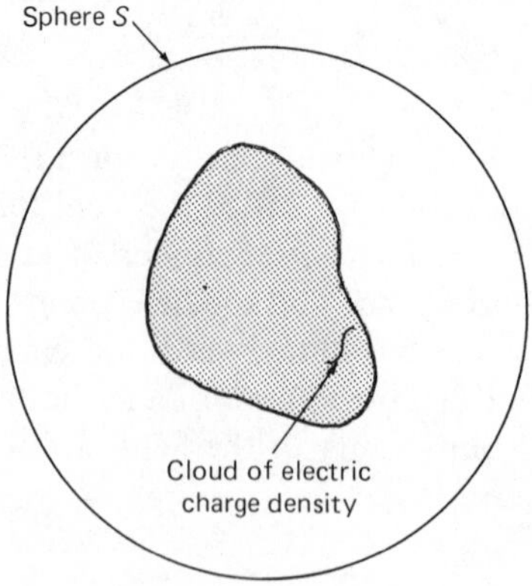

Fig. 2-3. Cloud of electric charge density.

by

$$\mathbf{E} = \int_V \frac{\rho \mathbf{a}_r}{\varepsilon_0 4\pi r^2} \, dv \tag{2.4.1-7}$$

and

$$\phi_e = \frac{1}{\varepsilon_0} \int_V \frac{\rho}{4\pi r} \, dv \tag{2.4.1-8}$$

where V is the volume enclosed by sphere S, and r is the distance between the field point $\mathbf{r}_f$ and the source point $\mathbf{r}_s$ (at which ρ is taken).

2.4.2. Biot-Savart Law

The Biot-Savart law relates to the static magnetic field in much the same way that Coulomb's law relates to the static electric field. That is, it expresses the static magnetic field as a summation of elementary sources in much the same way that Coulomb's law expresses the static electric field as a summation over elementary sources. Specifically, the Biot-Savart law expresses the static magnetic field as a summation over elementary current sources. In present-day notation, it is given as (Ref. 1, p. 221)

$$\mathbf{H}_{bs} = \int \frac{I \, \mathbf{dl} \times \mathbf{a}_r}{4\pi r^2} \tag{2.4.2-1}$$

In this equation, **dl** is an infinitesimal vector and I is the current over its length; that is, I **dl** is an infinitesimal *current element*. The unit radius vector, $\mathbf{a}_r$ goes *from* the source point (at which I is located) to the field point (at which **H** is taken). Equation (2.4.2-1) presumes a free-space universe (free of permeable materials and permanent magnets), and the integral is taken throughout all of space.

Suppose we have a universe that contains a volume current density, **J**. Suppose, also, that the medium of that universe is free-space throughout. Since the current element I **dl** in Equation (2.4.2-1) can be taken as small as we wish, we can derive from this equation an expression that gives the static magnetic field in this universe. This equation is

$$\mathbf{H}_{bs} = \int \frac{\mathbf{J} \times \mathbf{a}_r}{4\pi r^2} \, dv \tag{2.4.2-2}$$

where the integral is a volume integral over all of space.

By means of certain vector manipulations, Equation (2.4.2-2) can be put into a form that is more convenient for certain applications. We have

$$\frac{\mathbf{a}_r}{r^2} = -\nabla\left(\frac{1}{r}\right) = \nabla_s\left(\frac{1}{r}\right) \tag{2.4.2-3}$$

In Equation (2.4.2-3), the symbol ∇ means the gradient by differentiation with respect to the *field point*, while ∇_s means the gradient by differentiation with respect to the *source point*. Since $\mathbf{J}$ is taken at the source point, we have that

$$\nabla \times \left(\frac{\mathbf{J}}{r}\right) = \mathbf{J} \times \nabla\left(\frac{1}{r}\right) \tag{2.4.2-4}$$

From Equations (2.4.2-3) and (2.4.2-4), we have

$$\mathbf{J} \times \left(\frac{\mathbf{a}_r}{r^2}\right) = -\mathbf{J} \times \nabla\left(\frac{1}{r}\right) = -\nabla \times \left(\frac{\mathbf{J}}{r}\right) \tag{2.4.2-5}$$

From Equations (2.4.2-2) and (2.4.2-5), we have

$$\mathbf{H}_{bs} = -\frac{1}{4\pi}\nabla \times \int \frac{\mathbf{J}}{r}\,dv \tag{2.4.2-6}$$

From Equation (2.4.2-6), $\mathbf{H}_{bs}$ is the curl of a vector; and as a result,

$$\nabla \cdot \mathbf{H}_{bs} = 0 \tag{2.4.2-7}$$

2.5. INTEGRAL REPRESENTATIONS FOR QUASI-STATIC FIELDS USING THE HELMHOLTZ THEOREM

This section uses the Helmholtz theorem to develop integral representations, using the quasi-static approximation, for the **B**, **H**, **D**, and **E** fields. These integral representations are particularly useful in exterior problems.

2.5.1. Helmholtz Theorem

Appendix A derives the Helmholtz theorem for both two-dimensional and three-dimensional problems. These derivations apply to the following:

1. Bounded regions.
2. The unbounded region comprising all of space.
3. An unbounded region from which certain bounded regions are excluded.

The field and, in fact, *all fields* are assumed to go to zero over the boundaries of the bounded regions referred to above. In an unbounded region—also referred to above—we assume that all sources and conductive, permeable, and dielectric materials are confined within some bounded region, within some closed surface, say, a sphere. Outside such a surface, there is only free space. Thus, the magnetization field, **M**, and the polarization field, **P**, are zero outside this surface.

Appendix A shows that a vector field, **F**, can be expressed in terms of its curl and its divergence, taken over the entire region. Specifically,

$$\mathbf{F} = \nabla \times \mathbf{C} - \nabla\phi \tag{2.5.1-1}$$

where, for a three-dimensional region,

$$\mathbf{C} = \int_V G_3(\mathbf{r}_f, \mathbf{r}_s)\nabla_s \times \mathbf{F}\, dv \tag{2.5.1-2}$$

and

$$\phi = \int_V G_3(\mathbf{r}_f, \mathbf{r}_s)\nabla_s \cdot \mathbf{F}\, dv \tag{2.5.1-3}$$

In Equations (2.5.1-2) and (2.5.1-3), V is the volume of the region (infinite in the case of the unbounded region.)

$G_3(\mathbf{r}_f, \mathbf{r}_s)$ is called the *Green's function for three-dimensional free space* (see Ref. 3, p. 50) and is given by

$$G_3(\mathbf{r}_f, \mathbf{r}_s) = \frac{1}{4\pi|\mathbf{r}_f - \mathbf{r}_s|} \tag{2.5.1-4}$$

where $\mathbf{r}_f$ is the *radius vector to the field point* (the point at which the field, ϕ or **C**, is taken) and $\mathbf{r}_s$ is the radius vector to the source point. For a two-dimensional region, we replace G_3 in Equations (2.5.1-2) and (2.5.1-3) by G_2, the *Green's function for two-dimensional free space*, given by (see Ref. 3, p. 50)

$$G_2(\mathbf{r}_f, \mathbf{r}_s) = -\frac{1}{2\pi}\ln|\mathbf{r}_f - \mathbf{r}_s| \tag{2.5.1-5}$$

These equations then become

$$\mathbf{C} = \int_A G_2(\mathbf{r}_f, \mathbf{r}_s)\nabla_s \times \mathbf{F}\, dv \tag{2.5.1-6}$$

and

$$\phi = \int_A G_2(\mathbf{r}_f, \mathbf{r}_s)\nabla_s \cdot \mathbf{F}\, dv \tag{2.5.1-7}$$

In Equations (2.5.1-6) and (2.5.1-7), A is the cross-sectional area of the two-dimensional field.

The value of the Helmholtz theorem is that the curl and divergence of the field we wish to formulate are readily obtainable from Maxwell's equations. Therefore, this theorem makes it easy to develop the desired integral representations. The theorem also shows us that each of these fields can be split into the sum of two uniquely defined components, a curl-free component and a divergence-free component.

To save space as well as to save the reader time, integral expressions are derived below only for a three-dimensional region. Equations (2.5.1-2) through (2.5.1-7) show that one can modify any three-dimensional integral expression to a two-dimensional expression simply by replacing G_3 by G_2 and integrating over the cross-sectioned area, A, instead of over the volume, V. For simplicity, we have deleted the arguments $(\mathbf{r}_f, \mathbf{r}_s)$ in the equations shown in the remainder of this section.

2.5.2. Integral Expressions for B Field and Definition of Magnetic Vector Potential, A

Taking the curl of Equation (2.3.2-2) and combining it with Equation (2.2.2-1) yields, for the quasi-static case,

$$\nabla_s \times \mathbf{B} = \mu_0(\mathbf{J} + \nabla_s \times \mathbf{M}) \tag{2.5.2-1}$$

When **B** is substituted for **F** in Equations (2.5.1-1), (2.5.1-2), and (2.5.1-3) and these equations are combined with Equations (2.2.1-3) (that the divergence of **B** is zero) and (2.5.2-1), the result is

$$\mathbf{B} = \mu_0 \nabla \times \int_V G_3(\mathbf{J} + \nabla_s \times \mathbf{M})\, dv \tag{2.5.2-2}$$

This equation suggests the definition of the quasi-static *magnetic vector potential*, **A**:

$$\mathbf{A} = \mu_0 \int_V G_3(\mathbf{J} + \nabla_s \times \mathbf{M})\, dv \tag{2.5.2-3}$$

From Equations (2.5.2-2) and (2.5.2-3), we have

$$\mathbf{B} = \nabla \times \mathbf{A} \tag{2.5.2-4}$$

and Appendix B shows that

$$\nabla \cdot \mathbf{A} = 0 \tag{2.5.2-5}$$

There are three alternative forms of Equation (2.5.2-3) that are useful in some computations. First, we can define an *equivalent current density*, $\mathbf{J}_e$, by

$$\mathbf{J}_e = \nabla \times \mathbf{M} \tag{2.5.2-6}$$

Combining this with Equation (2.5.2-3) yields

$$\mathbf{A} = \mu_0 \int_V G_3(\mathbf{J} + \mathbf{J}_e)\, dv \tag{2.5.2-7}$$

In addition, as shown in Appendix B,

$$\mathbf{A} = \mu_0 \int_V [G_3\mathbf{J} - \mathbf{M} \times \nabla_s G_3]\, dv \tag{2.5.2-8}$$

and finally, using Equation (2.4.2-3),

$$\mathbf{A} = \mu_0 \int_V \left[G_3\mathbf{J} - \mathbf{M} \times \frac{\mathbf{a}_r}{r} G_3 \right] dv \tag{2.5.2-9}$$

2.5.3. Integral Expression for H

From Equation (2.2.1-3) and (2.3.2-2), we have

$$\nabla_s \cdot \mathbf{M} = -\nabla_s \cdot \mathbf{H} \tag{2.5.3-1}$$

When H is substituted for F in Equations (2.5.1-1), (2.5.1-2), and (2.5.1-3), and Equations (2.2.2-1) and (2.5.3-1) are used, we have that

$$\mathbf{H} = \nabla \times \int_V G_3\mathbf{J}\, dv + \nabla \int_V G_3 \nabla_s \cdot \mathbf{M}\, dv \tag{2.5.3-2}$$

Notice that the first term on the right of Equation (2.5.3-2) has zero divergence and that the second term on the right of the equation has zero curl. It turns out to be convenient in a number of problems to think of these terms as different components of the magnetic field, with names of their own. Then

$$\mathbf{H}_a = \nabla \times \int_V G_3 \mathbf{J}\, dv \tag{2.5.3-3}$$

and

$$\mathbf{H}_d = \nabla \int_V G_3 \nabla_s \cdot \mathbf{M}\, dv \tag{2.5.3-4}$$

so that

$$\mathbf{H} = \mathbf{H}_a + \mathbf{H}_d \tag{2.5.3-5}$$

In Equations (2.5.3-3), (2.5.3-4), and (2.5.3-5), $\mathbf{H}_a$ is called the *applied* magnetic field and $\mathbf{H}_d$ is called the *demagnetizing* magnetic field.

Equations (2.4.2-6), (2.5.1-4), and (2.5.3-3) show that $\mathbf{H}_a$ is given by the same expression as the Biot-Savart law (for a universe without permeable materials). This is to be expected. It turns out to be convenient in certain magnetic field calculations that $\mathbf{H}_a$ is independent of the permeability of the medium.

Since $\mathbf{H}_d$ is curl free, it is expressed in terms of the *reduced magnetic scalar potential*, ϕ_m, by

$$\mathbf{H}_d = -\nabla \phi_m \tag{2.5.3-6}$$

From Equations (2.5.3-4) and (2.5.3-6), we have that

$$\phi_m = -\int_V G_3 \nabla_s \cdot \mathbf{M}\, dv \tag{2.5.3-7}$$

In regions in which $\mathbf{J}$ is zero, $\mathbf{H}$ is curl-free; and then $\mathbf{H}$ can be expressed as the gradient of a *total* magnetic scalar potential, ψ_m, so that

$$\mathbf{H} = -\nabla \psi_m \tag{2.5.3-8}$$

Again, there are alternative forms for the integral in Equation (2.5.3-7) that are useful in certain computations. We can say that

$$\nabla \cdot \mathbf{M} = \rho_m \tag{2.5.3-9}$$

where ρ_m is called the *equivalent magnetic charge density*, and then from

Equations (2.5.3-7) and (2.5.3-9), we have

$$\phi_m = -\int_V G_3 \rho_m \, dv \tag{2.5.3-10}$$

In addition, certain vector manipulations can be performed upon Equation (2.5.3-7) to put it into a more convenient form. By a vector identity, this equation becomes

$$\phi_m = \int_V \mathbf{M} \cdot \nabla_s G_3 \, dv - \int_V \nabla_s \cdot (G_3 \mathbf{M}) \, dv \tag{2.5.3-11}$$

Applying the divergence theorem to the second integral in Equation (2.5.3-11) yields

$$\phi_m = \int_V \mathbf{M} \cdot \nabla_s G_3 \, dv - \int_S (G_3 \mathbf{M}) \cdot \mathbf{da}$$

and since $\mathbf{M} = 0$ over S, we have

$$\phi_m = \int_V \mathbf{M} \cdot \nabla_s G_3 \, dv \tag{2.5.3-12}$$

2.5.4. Integral Expression for E

Taking the divergence of Equation (2.3.1-3) and combining it with Equation (2.2.1-4) yields

$$\nabla \cdot \mathbf{E} = \frac{1}{\varepsilon_0}(\rho - \nabla \cdot \mathbf{P}) \tag{2.5.4-1}$$

Combining Equations (2.2.1-2) and (2.5.2-4) yields

$$\nabla \times \left(\mathbf{E} + \frac{\partial \mathbf{A}}{\partial t} \right) = 0 \tag{2.5.4-2}$$

When $\mathbf{F}$ in Equations (2.5.1-1), (2.5.1-2), and (2.5.1-3) is replaced by the expression $\mathbf{E} + \partial \mathbf{A}/\partial t$ and Equations (2.5.2-5), (2.5.4-1), and (2.5.4-2) are used, we have

$$\mathbf{E} + \frac{\partial \mathbf{A}}{\partial t} = \frac{1}{\varepsilon_0} \nabla \int_V G_3 (\nabla \cdot \mathbf{P} - \rho) \, dv \tag{2.5.4-3}$$

As with the magnetic field, it is convenient to divide the electric field, **E**, into the *applied* electric field, $\mathbf{E}_a$, and the *depolarizing* electric field, $\mathbf{E}_d$, so that

$$\mathbf{E} = \mathbf{E}_a + \mathbf{E}_d \tag{2.5.4-4}$$

$$\nabla \cdot \mathbf{E}_a = 0 \tag{2.5.4-5}$$

$$\nabla \times \mathbf{E}_d = 0 \tag{2.5.4-6}$$

From Equations (2.5.4-3) through (2.5.4-6), we have

$$\mathbf{E}_a = -\frac{\partial \mathbf{A}}{\partial t} \tag{2.5.4-7}$$

$$\mathbf{E}_d = \frac{1}{\varepsilon_0} \nabla \int_V G_3(\nabla \cdot \mathbf{P} - \rho)\, dv \tag{2.5.4-8}$$

As with the magnetic field, we define a *reduced electric scalar potential,* ϕ_e, by

$$\mathbf{E}_d = -\nabla \phi_e \tag{2.5.4-9}$$

From Equations (2.5.4-8) and (2.5.4-9), then,

$$\phi_e = \frac{1}{\varepsilon_0} \int_V G_3(\rho - \nabla \cdot \mathbf{P})\, dv \tag{2.5.4-10}$$

Again, in those regions that do not contain a changing magnetic field, the curl of **E** is zero, and we can define a *total electric scalar potential,* ψ_e, by

$$\mathbf{E} = -\nabla \psi_e \tag{2.5.4-11}$$

There are alternative useful forms of Equation (3.5.4-10). First, we can define an equivalent electric charge density, ρ_e, by

$$\rho_e = -\nabla \cdot \mathbf{P} \tag{2.5.4-12}$$

so that Equation (3.5.4-10) becomes

$$\phi_e = \frac{1}{\varepsilon_0} \int_V G_3(\rho + \rho_e)\, dv \tag{2.5.4-13}$$

Also, in much the same way that Equation (2.5.3-12) was derived, we have

$$\phi_e = \frac{1}{\varepsilon_0}\int_V [G_3\rho + \mathbf{P}\cdot\nabla G_3]\,dv \tag{2.5.4-14}$$

and

$$\phi_e = \frac{1}{\varepsilon_0}\int_V G_3\left(\rho + \mathbf{P}\cdot\frac{\mathbf{a}_r}{r}\right)dv \tag{2.5.4-15}$$

Notice that when ρ_e goes to zero in Equation (2.5.4-13), then ϕ_e reduces to Equation (2.4.1-8) (derived by Coulomb's law), as one would expect. Notice, further, from Equation (2.5.4-13) that ϕ_e can be considered to result simply from the spatial configurations of ρ and ρ_e. Similarly, from Equations (2.5.2-7) and (2.5.4-7), we see that $\mathbf{E}_a$ can be considered to result simply from the spatial configurations of $\mathbf{J}$ and $\mathbf{J}_e$.

2.5.5. Integral Expression for D

Taking the curl of Equation (2.3.1-3) and combining it with Equation (2.5.4-2) yields

$$\nabla \times \left(\mathbf{D} + \varepsilon_0\frac{\partial \mathbf{A}}{\partial t}\right) = \nabla \times \mathbf{P} \tag{2.5.5-1}$$

When $\mathbf{F}$ in Equations (2.5.1-1), (2.5.1-2), and (2.5.1-3) is replaced by the expression

$$\mathbf{D} + \varepsilon_0\frac{\partial \mathbf{A}}{\partial t}$$

and Equations (2.2.1-4), (2.5.2-5), and (2.5.5-1) are used, we have

$$\mathbf{D} + \varepsilon_0\frac{\partial \mathbf{A}}{\partial t} = \nabla \times \int_V G_3\nabla_s \times \mathbf{P}\,dv - \nabla\int_V G_3\rho\,dv \tag{2.5.5-2}$$

It is sometimes convenient to define a fictitious "magnetic current density," $\mathbf{J}_m$, given by

$$\mathbf{J}_m = \nabla \times \mathbf{P} \tag{2.5.5-3}$$

so that Equation (2.5.5-2) becomes

$$\mathbf{D} + \varepsilon_0\frac{\partial \mathbf{A}}{\partial t} = \int_V G_3\mathbf{J}_m\,dv - \nabla\int_V G_3\rho\,dv \tag{2.5.5-4}$$

We have seen that, in a free-space environment, the equivalent electric current density, $\mathbf{J}_e$, would contribute to the curl of **B** the same way that the curl of **M** would in a real permeable environment. In addition, the equivalent magnetic current density, $\mathbf{J}_m$, contributes to the curl of **D** in the same way that the curl of **P** does in a real dielectric medium. If, in Appendix B, Part 2, we substitute **P** for **M**, we obtain, from Equation (2.5.5-2),

$$\mathbf{D} + \varepsilon_0 \frac{\partial \mathbf{A}}{\partial t} = -\nabla \times \int_V \mathbf{P} \times \nabla_s G_3 \, dv - \nabla \int_V G_3 \rho \, dv \qquad (2.5.5\text{-}5)$$

and finally, using Equation (2.4.2-3),

$$\mathbf{D} + \varepsilon_0 \frac{\partial \mathbf{A}}{\partial t} = -\nabla \times \int_V \mathbf{P} G_3 \times \frac{\mathbf{a}_r}{r} \, dv - \nabla \int_V G_3 \rho \, dv \qquad (2.5.5\text{-}6)$$

2.6. EQUIVALENT CONFIGURATIONS

Equation (2.3.1-5) shows that at any point at which there is a dielectric material (and the electric susceptibility, χ_e, is positive) a polarization field, **P**, can exist. Similarly, at any point at which there is a permeable material (and the magnetic susceptibility, χ_m, is positive), a magnetization field, **M**, can exist. As shown in the previous section, to the extent that **M** and **P** have nonzero divergences and curls, they affect the **B**, **H**, **E**, and **D** fields. Also as shown in the previous section, the curls and divergences of **M** and **P** can be represented by *equivalent electric and magnetic current densities and charge densities.* We show below that certain of these *equivalent* charges and current configurations, in conjunction with the *real* charge and current configurations, can act in a free-space medium to produce the same fields as one would obtain in the real configuration, in which real charges and currents act in conjunction with whatever dielectric and permeable materials happen to be present.

From the Helmholtz theorem in Section 2.5.4, we see that, in the configuration to which it applies, the field is uniquely defined by its curl and its divergence. Therefore, all we have to do to create an equivalent field is to create a pattern of equivalent currents and/or equivalent charges such that the equivalent field and the real field have the same divergence and the same curl.

For the **B** field, we see from Equation (2.5.2-1) that a currently density $\mathbf{J} + \mathbf{J}_e$ with $\mathbf{J}_e$ given by Equation (2.5.2-6) will produce, in free space, an equivalent field that has the same curl as the real field and is therefore identical to the real field. (Both the real and equivalent fields have zero divergence.)

For the **H** field, we see from Equations (2.5.3-3) and (2.5.3-10) that a current density, **J**, and an equivalent magnetic charge density, ρ_m, given by Equation (2.5.3-9), will produce an equivalent field with the same divergence and curl as the real field. Therefore, this equivalent field will be the same as the real field throughout the region.

It was shown above that a current density, $\mathbf{J} + \mathbf{J}_e$ will, in free space, produce a **B** field equivalent to the **B** field in the real medium. With Equations (2.5.2-4) and (2.5.4-7), then, we see that this **B** field will cause the equivalent **E** field in free space, to have the same curl as the **E** field in the real medium. From Equations (2.5.4-1) and (2.5.4-12), we have that

$$\nabla \cdot \mathbf{E} = \frac{1}{\varepsilon_0}(\rho + \rho_e) \tag{2.6-1}$$

This equation shows us that a charge density of $\rho + \rho_e$ will cause the equivalent **E** field, in free space, to have the correct divergence. We see then that a charge density of $\rho + \rho_e$ and a current density of $\mathbf{J} + \mathbf{J}_e$ will, in a free space, produce an equivalent **E** field identical to the **E** field in the real medium.

From Equations (2.3.1-3) and (2.5.5-3), we have

$$\nabla \times \mathbf{D} = \varepsilon_0 \nabla \times \mathbf{E} + \mathbf{J}_m \tag{2.6-2}$$

We have shown that, in a free-space, current density $\mathbf{J} + \mathbf{J}_e$ yields the same **B** field and the curl of **E** as in a real permeable medium. Furthermore, in free space, the equivalent magnetic current density, $\mathbf{J}_m$, contributes to the curl of **D** in the same way that the curl of **P** does in a real dielectric medium. Then, from Equation (2.5.5-1), we see that, in free space, the current density $\mathbf{J} + \mathbf{J}_e + \mathbf{J}_m$ will produce the same curl of **D** as would occur in a real permeable *and* dielectric medium. From Equation (2.2.1-4) we have that the divergence of **D** is equal to the electric charge density, ρ. From these facts, we have that, in free space, the current density $\mathbf{J} + \mathbf{J}_e + \mathbf{J}_m$ and the charge density, ρ, will produce the same **D** field as in the real permeable and dielectric medium.

2.7. STEADY-STATE DYNAMIC PROBLEMS AND PHASOR FIELD REPRESENTATIONS

2.7.1. Steady-State Dynamic Problems

In this book we define a *steady-state dynamic problem* (or, for short, a *steady-state problem*) as a problem in which all fields vary sinusoidally with time, at all points in the region, at the same frequency. Clearly, the medium must be linear for a steady-state dynamic problem. Otherwise, certain fields, at certain points in the medium, would have nonsinusoidal time variations. As examples, calculation of fields from antennas, waveguides, and resonators and in magnetic cores, under the influence of eddy currents, are steady-state problems.

2.7.2. Phasor Field Representations

The fact that all fields vary sinusoidally at the same frequency in a steady-state problem permits great simplification in the formulation and representation of these fields. For example, consider the x component of the magnetic field, H_x, at some point in the region. It is given by

$$H_x = H_{x0} \cos(\omega t + \theta) \tag{2.7.2-1}$$

In this equation, ω is the frequency in radians per second, t is the time in seconds, θ is the phase angle of H_x at this point, and H_{x0} is simply the magnitude of H_x at this point. From this equation, we have

$$H_x = R_e[H_{x0} e^{j(\omega t + \theta)}] \tag{2.7.2-2}$$

In this equation, the function R_e yields the real part of its argument. The *phasor* representation of H_x, $\hat{H}_x$, is given by

$$\hat{H}_x = H_{x_0} e^{j\theta} \tag{2.7.2-3}$$

so that, with Equations (2.7.2-2) and (2.7.2-3),

$$H_x = R_e[\hat{H}_x e^{j\omega t}] \tag{2.7.2-4}$$

Notice that $\hat{H}_x$ is a *complex number* which is a function *only* of spatial coordinates (and not of time). Equation (2.7.2-4) exemplifies the general relationship between a field and its phasor representation. Notice that H_x is a *linear* function of $\hat{H}_x$. Furthermore, since, for any $\hat{H}_x$, there corresponds only one H_x, we have that $\hat{H}_x$ is a linear function of H_x. Equation (2.7.2-4) can be extended to vector fields; for example,

$$\mathbf{H} = R_e[\hat{\mathbf{H}} e^{j\omega t}] \tag{2.7.2-5}$$

where $\hat{\mathbf{H}}$ is the phasor representation of $\mathbf{H}$.

Using Equation (2.7.2-5), we can develop phasor equivalents to the field equations. For example, from Equation (2.2.1-1), we have

$$R_e[\nabla \times \hat{\mathbf{H}} e^{j\omega t}] = R_e[\hat{\mathbf{J}} e^{j\omega t}] + \frac{\partial}{\partial t} R_e[\hat{\mathbf{D}} e^{j\omega t}] \tag{2.7.2-6}$$

And

$$\frac{\partial}{\partial t} R_e[\hat{\mathbf{D}} e^{j\omega t}] = R_e[j\omega \hat{\mathbf{D}} e^{j\omega t}] \tag{2.7.2-7}$$

From Equations (2.7.2-6) and (2.7.2-7), we have

$$\nabla \times \hat{\mathbf{H}} = \hat{\mathbf{J}} + j\omega \hat{\mathbf{D}} \tag{2.7.2-8}$$

It is evident that the phasor equivalents of all field equations can be obtained easily, using a similar approach.

2.8. CONTINUITY CONDITIONS OF FIELDS AT A MEDIUM DISCONTINUITY

In our configurations, we frequently encounter surfaces, or interfaces, across which the medium varies discontinuously in permittivity, permeability, or conductivity. These discontinuities have a profound effect upon the fields and the equations and algorithms we use to compute the fields. It is essential to understand these effects very early in the computational process.

Continuity conditions are discussed below for the electric field, the magnetic field, the electric and magnetic scalar potentials, and the magnetic vector potential. Each vector field is decomposed into a component tangential to the interface, subscripted t, and a component normal to the interface, subscripted n. The permittivity, permeability, and conductivity are all assumed to be isotropic. That is, **D** and **J** have the same direction as **E**, and **B** has the same direction as **H**. All fields and media parameters on one side of the interface have the subscript 1, and those on the other side of the interface have the subscript 2.

2.8.1. Continuity of Electric Field

Since, from Equation (2.2.1-2), the curl of the electric field is finite, we can show that

$$\mathbf{E}_{t1} = \mathbf{E}_{t2} \tag{2.8.1-1}$$

Let $\mathbf{J}^t$ be the *total* current density, that is, the sum of the conduction current density and the displacement current density, so that

$$\mathbf{J}^t = \mathbf{J} + \frac{\partial \mathbf{D}}{\partial t} \tag{2.8.1-2}$$

Since the medium is isotropic in conductivity and permittivity, we have, from Equation (2.8.1-2),

$$\mathbf{J}' = \sigma \mathbf{E} + \varepsilon \frac{\partial \mathbf{E}}{\partial t} \tag{2.8.1-3}$$

where σ is the conductivity. The phasor equivalent of Equation (2.8.1-3) (see Section 2.7.2) is

$$\hat{\mathbf{J}}' = (\sigma + j\omega\varepsilon)\hat{\mathbf{E}} \tag{2.8.1-4}$$

From Equations (2.8.1-2) and (2.2.1-1), we have

$$\nabla \times \mathbf{H} = \mathbf{J}' \tag{2.8.1-5}$$

Taking the divergence of Equation (2.8.1-5) gives

$$\nabla \cdot \mathbf{J}' = 0 \tag{2.8.1-6}$$

from which

$$\mathbf{J}'_{n1} = \mathbf{J}'_{n2} \tag{2.8.1-7}$$

The continuity equation is obtained by combining Equations (2.8.1-4) and (2.8.1-7):

$$(\sigma_1 + j\omega\varepsilon_1)\hat{\mathbf{E}}_{n1} = (\sigma_2 + j\omega\varepsilon_2)\hat{\mathbf{E}}_{n2} \tag{2.8.1-8}$$

In practical application, the continuity equation, Equation (2.8.1-8), can almost always be simplified. This is because, for all good copper, iron, aluminum,

$$\sigma \gg \omega\varepsilon \tag{2.8.1-9}$$

for all reasonable permittivities and all frequencies up to X-rays. Using the inequality expressed in Equation (2.8.1-9), we can derive simpler continuity equations for the special case of the conductivity. If either σ_1 or σ_2 is nonzero, we have, from Equation (2.8.1-8),

$$\sigma_1 \hat{\mathbf{E}}_{n1} = \sigma_2 \hat{\mathbf{E}}_{n2} \tag{2.8.1-10}$$

If both σ_1 and σ_2 are zero, we have

$$\varepsilon_1 \hat{\mathbf{E}}_{n1} = \varepsilon_2 \hat{\mathbf{E}}_{n2} \tag{2.8.1-11}$$

If $\mathbf{E}_1$ and $\mathbf{E}_2$ were linearly polarized, then one could define the angles of incidence the $\mathbf{E}_1$ and $\mathbf{E}_2$ make with the interface surface. The condition that $\mathbf{E}_1$ is linearly polarized *requires* that $\mathbf{E}_{n1}$ and $\mathbf{E}_{t1}$ be in phase, and that $\mathbf{E}_{n2}$ and $\mathbf{E}_{t2}$ be in phase. But this is not necessarily true. Therefore, in general, the electric field in the dynamic steady-state case is elliptically polarized, and angles of incidence cannot be defined.

In the static case, assuming no free charge on the interface, we have at that interface, from Equation (2.2.1-4),

$$\nabla \cdot \mathbf{D} = 0$$

and from this that

$$\mathbf{D}_{n1} = \mathbf{D}_{n2} \tag{2.8.1-12}$$

and that

$$\varepsilon_1 \mathbf{E}_{n1} = \varepsilon_2 \mathbf{E}_{n2} \tag{2.8.1-13}$$

The angle of incidence, θ, is defined as the angle that the field makes with a normal to the interface. From this we have

$$E_n = E \cos \theta \tag{2.8.1-14}$$

and

$$E_t = E \sin \theta \tag{2.8.1-15}$$

and from these equations that

$$\tan \theta = \frac{E_t}{E_n} \tag{2.8.1-16}$$

From Equations (2.8.1-1) and (2.8.1-13), we have that

$$\varepsilon_1 \frac{E_{n1}}{E_{t1}} = \varepsilon_2 \frac{E_{n2}}{E_{t2}}$$

and with Equation (2.8.1-16),

$$\frac{\varepsilon_1}{\tan\theta_1} = \frac{\varepsilon_2}{\tan\theta_2} \qquad (2.8.1\text{-}17)$$

where θ_1 and θ_2 are the angles of incidence if the electric field is on the medium #1 side and the medium #2 side of the interface.

2.8.2. Continuity of Magnetic Field

From Equation (2.2.1-1), the curl of the magnetic field is finite, with the result that

$$H_{t1} = H_{t2} \qquad (2.8.2\text{-}1)$$

From Equation (2.2.1-3), the divergence of the magnetic flux density, B, is finite, with the result that

$$B_{n1} = \mu_1 H_{n1} = \mu_2 H_{n2} = B_{n2} \qquad (2.8.2\text{-}2)$$

Again, in the static case, the angle of incidence, θ, is defined as the angle that the magnetic field makes with a normal to the interface, so that

$$H_n = H\cos\theta \qquad (2.8.2\text{-}3)$$

and

$$H_t = H\sin\theta \qquad (2.8.2\text{-}4)$$

and from these equations that

$$\tan\theta = \frac{H_t}{H_n} \qquad (2.8.2\text{-}5)$$

From Equations (2.8.2-1) and (2.8.2-2), we have

$$\mu_1\frac{H_{n1}}{H_{t1}} = \mu_2\frac{H_{n2}}{H_{t2}}$$

and with Equation (2.8.2-5),

$$\frac{\mu_1}{\tan\theta_1} = \frac{\mu_2}{\tan\theta_2} \qquad (2.8.2\text{-}6)$$

where θ_1 and θ_2 are the angles of incidence of the electric field on the medium #1 side and the medium #2 side of the interface.

2.8.3. Continuity of Magnetic Vector Potential and Scalar Potential

From Equations (2.5.2-4) and (2.5.2-5), we see that both the curl and the divergence of the magnetic vector potential, **A**, are everywhere finite. From this we see that **A** is continuous over any interface.

From Equations (2.5.3-6) and (2.5.4-9), we see that the gradients of the magnetic scalar potential, ϕ_m, and the electric scalar potential, ϕ_e, are finite everywhere. Therefore, these scalar potentials are continuous across any interface.

REFERENCES

1. Elliott, Robert S. *Electromagnetics.* New York: McGraw-Hill Book Co., 1966.
2. Plonsey, Robert, and Collin, Robert E. *Electromagnetic Fields.* New York: McGraw-Hill Book Co., 1961.
3. Stakgold, Ivar. *Boundary Value Problems of Mathematical Physics,* Vol. II. New York: The Macmillan Company, 1967.

3
PROBLEM DEFINITION

3.1. INTRODUCTION

A very important part of the job of computing a field is that of defining the problem. In fact, whether the project succeeds or fails may depend upon the skill with which the problem is defined.

As a matter of practical necessity, the field problem that we solve numerically differs in certain ways from the actual, physical field problem. That is, the physical configuration in which the actual field exists, in terms of the boundaries, boundary conditions, media, etc., differ somewhat from the boundaries, boundary conditions, media, etc., that we assume for the field problem that we solve. In this book we call the actual field problem in the actual physical configuration the *real* field problem. The field problem that we solve numerically, with its approximate configuration, we call the *approximate* field problem. This book is concerned almost entirely with the approximate field problem. In fact, one can assume that any reference to a field problem is to the approximate field problem unless otherwise stated.

Throughout the remainder of this book, the terms *exact solution* and *approximate numerical solution* are used. These terms invariably mean the exact solution and the approximate numerical solution to the *approximate* field problem. We assume, then, that if our approximate field problem is similar enough to our real field problem, and if our approximate numerical solution is a good enough approximation to the exact solution of our approximate field problem, then our approximate solution will be a good enough approximation to the real field in the real field problem.

There are a number of properties of the physical configuration in which the field resides that have a considerable effect upon the field. These properties are discussed in the remaining sections of this chapter. It is not possible, in general, to assign a relative importance to these different properties. Nor is there a natural order in which they should be discussed, and no attempt has been made to put them in such an order. For many of the topics discussed in this book, there is a natural, logical, order for the discussion—that is, a natural starting point, a logical order in which the points of the topic are to be made, and a natural ending point. This is not so with problem definition.

A field problem is defined so as to meet the following objectives:

1. Duplicate as nearly as possible the actual physical environment of the fields (to promote the accuracy of the computed field).
2. Have a unique solution (for a deterministic problem).
3. Have one or more practical algorithms by which it can be solved.
4. Be numerically soluble within the constraints of manpower, time, computer facilities, and funds available.

In practice, finding a problem definition that fits all of the above objectives is frequently difficult or impossible. Then, for one or more of the physical properties given below, the definition that is used can become a compromise between the accurate modeling of reality and the limitations of a practical computation.

It is frequently time well spent to consider, for a given problem, a number of problem definitions. In this way, one can reasonably hope to arrive at the optimum compromise between computational accuracy and a practical computation.

3.2. FIELD PROBLEM DOMAINS, SOURCE PROBLEM DOMAINS, INTERIOR PROBLEMS, AND EXTERIOR PROBLEMS

In some of our problems, we include all of space (i.e., the universe) and consider an electric or magnetic field throughout all of space. But for most of our problems, certain regions, and the fields that reside in these regions, are excluded from the problem. In defining a problem, it is important to know, from the outset, what portions of the universe are included and what portions are excluded. To simplify this, we term that portion of space in which we compute the field to be the *domain of the field problem* or the *field problem domain.* Any field problem domain that does not comprise the entire universe is bounded by *boundaries.* We say that such a problem is a *bounded domain problem.* Figure 3-1 shows examples of certain problem domains.

Figure 3-1a shows the field problem domain of an *interior* problem. The field problem domain is enclosed within a boundary. All space outside this boundary is excluded from the field problem domain. Figure 3-1b shows the field problem domain *exterior* problem. Here, the field problem domain extends in all directions to infinity. In the case shown, two regions that are surrounded by the field problem domain are excluded from it. As we see in subsequent chapters, the algorithms we use for interior problems are usually quite different from those that we use for exterior problems. In most cases, exterior numerical field problems are more difficult than interior numerical field problems.

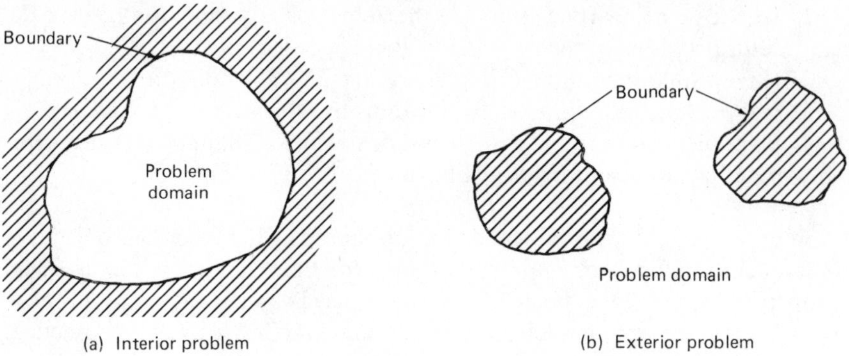

Fig. 3-1. Interior and exterior problem domains.

Suppose that the two boundary surfaces that surround the excluded regions in Figure 3-1b have distributed over them certain sources. These sources might, for example, be electric charges. Suppose, further, that the fields in the field problem domain of this figure can be written as integrals over these sources. We say, then, that these two boundary surfaces constitute a *source problem domain.* In the problems considered in this book, source problem domains can, in some cases, be surfaces, as in this case, or they can, in other problems, be three-dimensional regions or even lines.

Problem domains, their boundaries, and boundary conditions usually differ between real problems and approximate problems. As shown in Chapter 4, Section 4.4.1, the approximate problem domain is the union of the finite elements we use, and it is sometimes not practical for this union to coincide exactly with the real problem domain. In these cases, the approximate boundary differs slightly from the exact boundary, and therefore the approximate boundary conditions differ from the real boundary conditions. Throughout the remainder of this book, unless otherwise noted, the problem domains, boundaries, and boundary conditions used are the approximate ones.

3.3. IS THE PROBLEM STATIC, QUASI-STATIC, OR DYNAMIC?

This question was discussed briefly in Chapter 2, Section 2.2. As shown there, if the displacement current can be neglected, the quasi-static approximation holds. If, in addition, the effect of the time-rate of change of magnetic field upon the electric field can be neglected, the problem is a static one. If neither of these approximations holds, the problem is a fully dynamic one. Only static and quasi-static problems are considered in this book.

Usually, quasi-static problems are easier to solve than dynamic problems, and static problems are easier to solve than quasi-static problems. Since most numerical problems are difficult to solve at best, we would very much like our problem to be a static problem or, if not that, a quasi-static problem. Thus, we have a strong incentive for ascertaining which of these approximations, if any, hold for a given problem.

3.4. WHAT FIELD IS TO BE COMPUTED?

Sometimes we compute scalar fields (such as the scalar electric potential), and sometimes we compute vector fields (such as the electric or magnetic field). In general, we try very hard to define our problems in such a way that we compute a one-component field, since they are usually much easier to compute than two- or three-component fields. (By one-component field, we mean either a scalar potential or a single-component vector field.)

3.5. IS THE PROBLEM TWO-DIMENSIONAL OR THREE-DIMENSIONAL?

In a two-dimensional problem, we are, of course, really approximating a three-dimensional problem. Specifically, it is a problem in which we compute fields in a certain plane—the plane of computation. In the vicinity of this plane, the media and sources vary negligibly in a direction normal to the plane over a long distance either positively or negatively, in that normal direction. As a result, we can, to a reasonable degree of approximation, assume that the fields do not vary in that normal direction.

Two-dimensional problems are quite popular in numerical field computations. As shown in Chapters 4 and 5, two-dimensional field problems require simpler field formulations and simpler field approximations than do three-dimensional field problems. Furthermore, they are much less demanding of computer memory and computer time, and require simpler computer programs.

We use two-dimensional problems wherever the configuration of media and sources allows us to do so.

3.6. THE MEDIUM

The nature of the medium in conductivity, permittivity, and permeability is one of the most important properties of the problem. It determines whether the problem can be solved numerically, and, if so, how difficult that solution will be.

3.6.1. Is the Medium Isotropic?

If a medium is isotropic at a point, then it has the same behavior with respect to a field at a point regardless of the direction of that field. For example, if the permeability μ is isotropic, then, at any given point,

$$\mathbf{B} = \mu\mathbf{H} \tag{3.6.1-1}$$

with **H** and **B** being in the same direction, *regardless of what that direction is.* Any medium that does not have this property is called *nonisotropic.*

This book is limited to problems in which the permeability, permittivity, and conductivity are isotropic. Problems with nonisotropic media are extremely difficult—indeed practically impossible—in many configurations of the medium. Such problems are an important study in themselves.

3.6.2. Medium Linearity

If we say that, say, the permeability is linear in a problem, we mean that at *every* point within the region of computation, there is a linear relationship between **B** and **H**, **D** and **E**, and **J** and **E**. That is, at each point in this region, for example,

$$\mathbf{B} = \mu\mathbf{H}$$

where μ is a constant. If, at any point in this region, we do not have a linear relationship between **B** and **H**, so that μ in this equation is not constant at that point, we say that the permeability is *nonlinear*. Similar statements apply to permittivity and conductivity. If a problem is linear in permeability, permittivity, and conductivity, we say that the problem is linear. If the problem is nonlinear in permeability, permittivity, *or* conductivity, we say that the problem is nonlinear.

As we will see in the remaining chapters, linear problems are much easier than nonlinear problems. The fact that, in a linear problem, both the medium and Maxwell's equations are linear greatly simplifies our mathematics. It enables us to make relatively simple and straightforward use of such mathematical tools as calculus, vector analysis, and linear algebra. For nonlinear problems, these mathematical tools become more difficult to apply. For example, we find that most linear numerical field problems are solved by constructing and solving a linear system of equations. A nonlinear problem is solved by setting up and solving a nonlinear system. We find in practice that a nonlinear system requires not only much more computer time than a linear system but also more care on the part of the mathematician.

3.6.3. Medium Uniformity

In most static and quasi-static problems that we encounter in practice, the medium is nonuniform in permeability, permittivity, or conductivity. Furthermore, these nonuniformities usually have a very strong effect upon the fields. In fact, permeability, permittivity, or conductivity, or some combination of these, usually varies discontinuously over certain boundaries within the region. To discuss the typical medium nonuniformities in more definite terms, we can use these boundaries to divide up the field problem domain into a number of subdomains. That is, the medium is continuous in permeability, permittivity, and conductivity throughout each of these subdomains, but one or more of these properties is discontinuous across the subdomain boundaries. In practice, we find subdomains of two types.

1. Subdomains in which the permeability, permittivity, and conductivity are all uniform throughout.
2. Subdomains in which one or more of these properties varies continuously throughout.

In those problems in which we find a medium of the first type, we are careful to take advantage of that uniformity, since it invariably simplifies our computation.

3.7. BOUNDARY CONDITIONS

In a bounded domain problem, we choose the boundaries (and therefore the field problem domain) in such a way that we have some knowledge of the behavior of the field over these boundaries. We call this behavior of the field at the boundaries the *boundary conditions.* The definitions of these boundary conditions in mathematical terms are given below. Typically, our knowledge of these boundary conditions comes from knowing something of the medium or of any generators within the excluded regions. Then, since electric and magnetic fields must satisfy certain continuity conditions (see Section 2.8), we know that these boundary conditions must exert a strong effect upon the fields within the field problem domain. In fact, by the use of certain *uniqueness theorems,* we know that for certain problems, certain boundary conditions, in conjunction with Maxwell's equations (see Section 2.2), *uniquely prescribe* the fields within the field problem domain.

3.7.1. Scalar Boundary Conditions

We use scalar boundary conditions in most static problems.

The *Dirichlet* boundary condition prescribes the *value* of the scalar potential, γ, over a portion or all of a boundary and is given by

$$\gamma(\mathbf{r}_b) = g(\mathbf{r}_b) \tag{3.7.1-1}$$

In Equation (3.7.1-1), $\mathbf{r}_b$ is the radius vector to a point on the boundary and $g\ (\mathbf{r}_b)$ is, of course, the value of the prescribed scalar potential at that point. The *Neumann* boundary condition prescribes the *normal derivative* of the scalar and is written

$$\frac{\partial \gamma(\mathbf{r}_b)}{\partial n} = h(\mathbf{r}_b) \tag{3.7.1-2}$$

Finally, we have the *mixed* boundary condition, given by

$$\frac{\partial \gamma(\mathbf{r}_b)}{\partial n} + \sigma(\mathbf{r}_b)\gamma(\mathbf{r}_b) = h(\mathbf{r}_b) \tag{3.7.1-3}$$

Notice that if

$$\sigma(\mathbf{r}_b) = 0$$

then the mixed boundary condition reduces the Neumann boundary condition.

3.7.2. Vector Boundary Conditions

We use vector boundary conditions in both static and quasi-static problems. In some cases we prescribe the components of **E**, **H**, and **A** that are tangential to the boundary, that is, $\mathbf{E}_t$, $\mathbf{H}_t$ and, $\mathbf{A}_t$. In other cases, we prescribe the components of **B** and **D** that are normal to the boundary, that is $\mathbf{B}_n$ and $\mathbf{D}_n$.

4
LINEAR SPACES IN FIELD COMPUTATIONS

4.1. INTRODUCTION

The equations that govern field behavior are *linear* in the electric, magnetic, and related fields. In these equations, the operations that are performed on the fields, differentiations and integrations, are linear operations. Because of this linearity, linear algebra is a very useful tool in numerical field computation. It is useful in helping us develop good approximate representations for these fields that are suitable for use in numerical computations and in developing and understanding the functioning of the algorithms we use.

Any numerical field computation by the integral equation method or the finite element method will contain, among others, the following steps.

1. Decide upon the region over which the integral equation is to be solved, or the region over which a finite element solution will be obtained. As discussed in Chapter 3, this region is called the *approximate problem domain.*
2. Define an infinite-dimensional linear space, S, that has as elements functions having the approximate problem domain as a domain. (see Fig. 4-4-b) The exact field or source distribution over this problem domain is an element of S.
3. Set up a finite-dimensioned subspace of S, S_N, of dimension N. The space S_N then has, as an element, the approximate field or source distribution that is to be numerically computed.
4. Based upon space S_N and the field equations provided in Chapter 2, set up a system of N equations.
5. Solve this system of equations. This N-dimensional solution vector then identifies the element of S_N that is the numerically computed approximate field or source distribution.

While the above steps may seem artificial and abstract at this point, they take on very specific physical meaning in terms of the sample problems that are presented in Chapters 6 and 8.

The linear space S of step 2 above is simply an abstraction that is helpful in developing and understanding our algorithms.

On the other hand, it is quite important, in terms of making accurate field computations and minimizing required computer resources (such as central processor time and rapid-access memory), to do a good job in constructing the subspace S_N. Most of the remainder of this chapter is intended to aid in this construction, which is often, of practical necessity, done with the aid of a computer.

The construction of S_N requires the following steps:

1. Subdivision of the problem domain into finite elements. (See Figure 4-4.)
2. Choice of node points within the problem domain. A node point is a point in the problem domain at which the field or source distribution is computed.
3. Choice of shape functions. A shape function is a continuous function that is defined over a single finite element and that equals unity at one node point within the finite element and zero at all other node points within the finite element.
4. Definition of basis functions.

The order in which these steps are taken depends upon the nature of S_N. Section 4.2 gives a general discussion of the basis functions that we use.

There is a considerable advantage to adapting the configuration of finite elements to the geometry of the region in which the fields are to be computed (the problem domain), its conductors and blocks of permeable and dielectric materials. For this reason, a wide variety of finite element shapes, and shape functions and basis functions that correspond to these finite elements, have been discussed in the literature and used in practice. Certain of these finite elements, shape functions, and basis functions, in both two dimensions and three dimensions, and discussed below. In addition, references are given for discussion of these items in other works.

4.2. BASIS FUNCTIONS

A basis function, as it is used in field computations, has the following properties:

1. Each basis function has, as its domain, the entire problem domain.
2. Each basis function corresponds to just one node point.
3. Each basis function is nonzero over just those finite elements that contain its node point. That is, the basis function equals zero over all other finite elements. Another way to describe this property is to say that the basis function has *local support*. The union of finite elements over which the basis function can be nonzero is called its *support*.
4. The basis function has a value of unity at its node point.
5. The N basis functions are *linearly independent*. That is, no basis function equals a linear combination of the other basis functions.

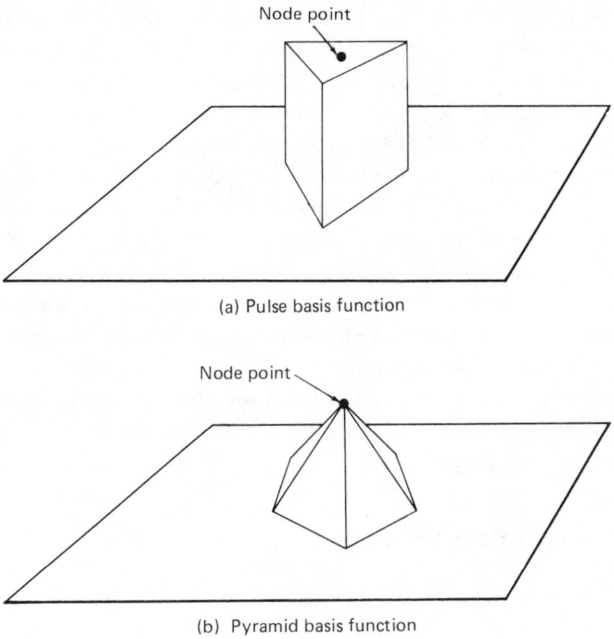

Fig. 4-1. Basis functions is two dimensions.

We represent the basis functions of S_N by $\boldsymbol{\alpha}_i$ for $1 \leqslant i \leqslant N$. Since these basis functions are linearly independent, any element of S_N (including the numerically computed field) can be represented as a linear combination of basis functions. That is, any element, $\boldsymbol{\beta}$, of S_N can be represented as

$$\boldsymbol{\beta} = \sum_{i=1}^{N} \beta_i \boldsymbol{\alpha}_i \tag{4.2-1}$$

where the β_i are real numbers.

There are two types of basis functions in common use, *pulse basis functions* and *pyramid basis functions*. Examples of pulse and pyramid basis functions for two-dimensional problem domains are shown in Figures 4-1a and 4-1b, respectively.

4.2.1. Pulse Basis Functions

Each pulse basis function corresponds to just one finite element. As shown in Figure 4-1a, the pulse basis function equals unity over its finite element and equals zero over the remainder of the problem domain. Typically, the pulse basis function's node point is at or near the center of the finite element.

Notice from Figure 4-1a that the pulse basis function is discontinuous over the boundary of its finite element. Its gradient is zero except on this boundary. Suppose we have a linear space, S_N, that uses pulse basis functions. Then, by Equation (4.2-1), each element of S_N is a linear combination of these basis functions. Any element of this linear space may be discontinuous at the finite element boundaries and will have a zero gradient at all points that are not on these boundaries. These properties make pulse basis functions unacceptable for applications in which each element must be globally continuous (continuous at all points in the problem domain) and in which the difference between the gradient of the exact field and the gradient of an element of S_N is to be minimized. As shown below, the finite element method is such an application. A pulse-type basis function can, however, be used to represent an approximate source function in the solution of an integral equation by the point-matching or collocation method. In this application, they have the advantage of relative simplicity (Ref. 1, pp. 503–505).

4.2.2. Pyramid Basis Functions

A pyramid basis function is characterized by the following:

1. It takes its maximum *only* at or near its node point.
2. It equals unity at its node point.
3. Its gradient is nonzero over only the finite elements that contain its node point (its support).
4. It is continuous over the entire problem domain.

These features are exemplified by Figure 4-1b. Since any element of S_N is a linear combination of its basis functions, then, if any S_N uses pyramid basis functions, then any element of S_N is continuous over the entire problem domain. Furthermore, the gradient of any element is, in general, nonzero. These properties makes pyramid basis functions useful in the finite element method.

The pyramid basis function is more complicated than the pulse basis function in two ways:

1. While the pulse basis function is nonzero over only one finite element, the pyramid basis function can be nonzero over more than one finite element, that is, over all finite elements that contain the basis function's node point, and constitute its support.
2. While, over each finite element, the gradient of the pulse basis function is zero, over certain finite elements (of its support) the gradient of the pyramid basis function is nonzero.

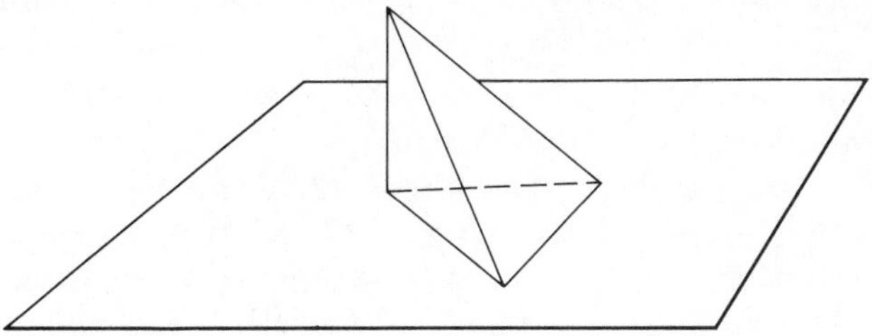

Fig. 4-2. Lowest order shape function over a triangle.

For these reasons, it is necessary, in the construction of a pyramid basis function, to think of it as consisting of one or more *shape functions*. Each of these shape functions is defined over one of the finite elements that contains the basis function's node point. Over each of these, the basis function equals the shape function. As an example, Figure 4-2 shows a linear shape function defined over a triangular finite element.

The construction of shape functions is a very important aspect of the development of the linear space, S_N, and, in fact, in the numerical computation of electric and magnetic fields. Furthermore, a wide variety of shape functions of different types are needed to fit the various geometrical configurations in which fields are computed. For these reasons, Sections 4.4, 4.5, and 4.6 are devoted to the construction of shape functions in two dimensions and three dimensions.

4.2.3. Lowest Order and Higher Order Pyramid Basis Functions

Higher order shape functions and basis functions and their node point structures occur in both two-dimensional and three-dimensional problem domains. However, those concepts are more easily explained and understood in two dimensions, and so that is what is done here.

In the examples of basis functions and shape functions that have been given thus far, there are node points *only* at the vertices of the finite elements. Shape functions over finite elements that have node points only at their vertices are called *lowest order shape functions*. An example of these is shown in Fig. 4-2. Pyramid basis functions that are constructed out of lowest order shape functions are called *lowest order basis functions*.

As shown in Figure 4-3, it is also possible to have node points midway along the sides of finite elements, or even in the interiors of finite elements. The shape functions over such finite elements are called *higher order shape functions* and the basis functions made up out of these are called *higher order basis functions*.

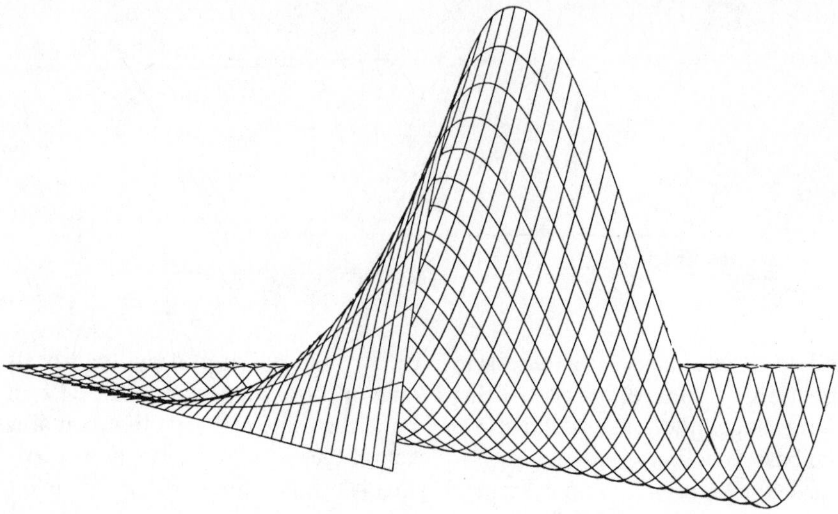

Fig. 4-3a. Higher order shape functions over a triangle. Node point midway along side of a triangle.

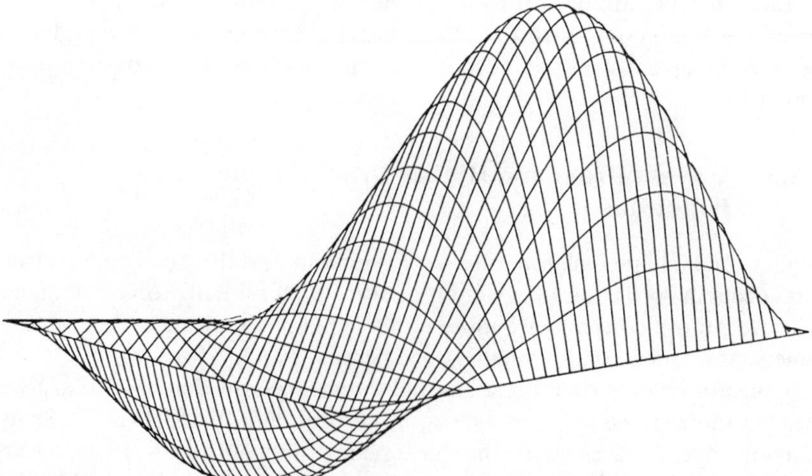

Fig. 4-3b. Higher order shape functions over a triangle. Node point in interior of a triangle.

That is, the shape functions shown in Figure 4-3 are higher order shape functions. There is a basis function that corresponds to each node point, whether the node point is at the vertex of a finite element, midway along its sides, or within its interior. In the basis function of Figure 4-1b, the node point is at a vertex that is shared by six triangles; so the basis function is nonzero over all six of these triangles. Alternatively, one could construct a basis function over two triangular adjacent finite elements by using shape functions of the type shown in Figure 4-3a. That is, the two shape functions share a common node point (that of the basis function) along their common side. Finally, there is a basis function that consists only of the shape function shown in Figure 4-3b, since the node point of that shape function is not shared by any other shape function.

4.3. SHAPE FUNCTIONS

The shape function has the following properties:

1. It has as its domain just one of the finite elements.
2. It is continuous over that finite element.
3. It equals unity at one of its node points.
4. It equals zero at all other node points within its finite element.

A complete discussion of shape functions is beyond the scope of this book. For shape functions that are not covered here, the reader is referred to Mitchell and Wait (Ref. 2, pp. 64–102). In addition, these authors discuss shape functions in one dimension as a help to the reader in getting a good "feel" for shape functions (Ref. 2, pp. 1–21).

One of the many differences among shape functions is the degree of continuity that they maintain at the finite element boundaries. If Lagrange interpolation is used (Ref. 2, p. 64), then an element of S_N is continuous at finite element boundaries, but the normal derivative of the element at these boundaries is generally discontinuous. If Hermite interpolation is used, *both* the element *and* its normal derivative are continuous at the finite element boundaries. Hermite interpolation is usually more complicated than Lagrange interpolation. The degree of continuity needed in elements of S_N is one less than the highest order of differentiation that occurs in the bilinear forms in which those elements are used (Ref. 3, p. 52). As we will see later in the problems to which this book is addressed, this highest order of differentiation can always be reduced to unity (if necessary, by integration by parts). Therefore, continuity of the gradient (or the normal derivative) is not needed in the problems addressed in this book. That is, Hermite interpolation is not needed. To avoid unnecessary complexity, we use only Lagrange shape functions here.

4.4. FINITE ELEMENTS AND SHAPE FUNCTIONS OF GLOBAL COORDINATES IN TWO-DIMENSIONAL PROBLEM DOMAINS

This section and the next section deal in detail with the subdivision of a two-dimensional problem domain into finite elements and the construction of shape functions over those elements. This problem domain can be either a region in two-dimensional space (an area) or a surface (that can be curved) in three-dimensional space.

4.4.1. Subdivision of a Two-dimensional Problem Domain

There are, of course, any number of ways into which a curved surface or plane segment can be subdivided into finite elements. The two most popular are rectangles and triangles, as shown in Figures 4-4a and 4-4b. Subdivision into rectangles has the advantage of simplicity if the problem domain is shaped so that it can be covered easily by rectangles. Subdivision into triangles has the advantage of fitting any irregularly shaped problem domain. Furthermore, one can cover the problem domain by any irregular distribution of node points and then connect these node points with straight-line segments in such a way as to form triangles.

Figure 4-4b shows both the real boundary and the approximate boundary. The question arises, why did we not use finite elements along the boundary whose sides exactly fit the real boundary, so that the real boundary and the approximate boundary would be the same? The answer lies in the shape functions that we use, since the domain of each shape function is its finite element. We find, in practice, that we can define shape functions over triangles and over figures with certain simple curved sides by expressions of only moderate complexity. But a shape function defined over a finite element with

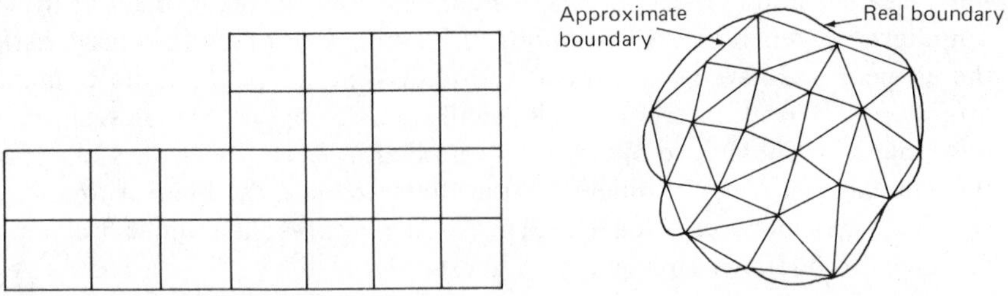

(a) Subdivision of a two-dimensional problem domain into rectangles

(b) Subdivision of a two-dimensional problem domain into triangles

Fig. 4-4. Subdivision of two-dimensional problem domains.

a side of *any prescribed* shape can require an expression that is too complex for our purposes. For these reasons, the lesser of the two evils is to accept an approximate boundary that is slightly different from the real boundary, as shown in Figure 4-4b.

4.4.2. Linear Shape Function over a Triangular Finite Element

We proceed to express the linear or lowest order shape function over a triangular finite element, $p_i(x, y)$, as shown in Figure 4-2. Since $p_i(x, y)$ is linear, its triangle has node points only at the vertices. Here, the superscript of p is the degree of the shape function and the subscript of p is the index of the vertex of the triangle at which the shape function equals unity. That is,

$$p_i^1(x_j, y_j) = \delta_{ij}, \qquad 1 \leqslant i, j \leqslant 3 \tag{4.4.2-1}$$

where (x_j, y_j) for $1 \leqslant j \leqslant 3$ are the triangle vertices. In Equation (4.4.2-1), the symbol δ_{ij} is the *Kronecker delta*, defined as follows (Ref. 7):

$$\delta_{ij} = 0 \qquad \text{for} \qquad i \neq j$$

$$\delta_{ij} = 1 \qquad \text{for} \qquad i = j$$

Since $p_i(x, y)$ varies linearly with x and y, we have that

$$p_i^1(x, y) = a_{i1} + a_{i2}x + a_{i3}y, \qquad 1 \leqslant i \leqslant 3 \tag{4.4.2-2}$$

When Equations (4.4.2-1) and (4.4.2-2) are combined, we have the linear systems

$$a_{i1} + a_{i2}x_j + a_{i3}y_j = \delta_{ij}, \qquad 1 \leqslant i, j \leqslant 3 \tag{4.4.2-3}$$

There are nine Equations (4.4.2-3) that constitute three third-order linear systems, that is, a linear system for $i = 1$, a linear system for $i = 2$, and a linear system for $i = 3$. When we solve these linear systems by Cramer's rule (Ref. 4, p. 84), and we notice that the area A of the triangle is given by

$$A = \tfrac{1}{2}\det \begin{vmatrix} 1 & x_1 & y_1 \\ 1 & x_2 & y_2 \\ 1 & x_3 & y_3 \end{vmatrix} \tag{4.4.2-4}$$

we find that

$$p_1^1(x, y) = \frac{x_2 y_3 - x_3 y_2 + (y_2 - y_3)x + (x_3 - x_2)y}{2A} \tag{4.4.2-5}$$

$$p_2^1(x, y) = \frac{x_3 y_1 - x_1 y_3 + (y_3 - y_1)x + (x_1 - x_3)y}{2A} \tag{4.4.2-6}$$

$$p_3^1(x, y) = \frac{x_1 y_2 - x_2 y_1 + (y_1 - y_2)x + (x_2 - x_1)y}{2A} \tag{4.4.2-7}$$

4.4.3. Triangular Area Coordinates

Triangular area coordinates are much more convenient than Cartesian coordinates in certain applications, such as higher order shape functions over triangles and numerical integration over triangles (Ref. 5, pp. 116–117). They are also referred to as Mobius coordinates, barycentric coordinates (Ref. 6, pp. 298–300), or homogeneous coordinates (Ref. 3, p. 129). Each point, P, in a triangle has three area coordinates (which are dependent). Each of these area coordinates can be defined either in terms of distances or in terms of areas. The definitions given below follow those of Vichnevetsky (Ref. 6, pp. 298–300). Consider the triangle in Figure 4-5, and the point P in that triangle. The area coordinates Λ_1, Λ_2, and Λ_3 can be defined as

$$\Lambda_1 = \frac{\text{distance of } P \text{ to side 1}}{\text{distance of vertex 1 to side 1}} \tag{4.4.3-1}$$

$$\Lambda_2 = \frac{\text{distance of } P \text{ to side 2}}{\text{distance of vertex 2 to side 2}} \tag{4.4.3-2}$$

$$\Lambda_3 = \frac{\text{distance of } P \text{ to side 3}}{\text{distance of vertex 3 to side 3}} \tag{4.4.3-3}$$

From these definitions, we can see that Λ_1, Λ_2, Λ_3 can also be expressed as

$$\Lambda_1 = \frac{\text{area of subtriangle 1}}{\text{area of total triangle}} \tag{4.4.3-4}$$

$$\Lambda_2 = \frac{\text{area of subtriangle 2}}{\text{area of total triangle}} \tag{4.4.3-5}$$

$$\Lambda_3 = \frac{\text{area of subtriangle 3}}{\text{area of total triangle}} \tag{4.4.3-6}$$

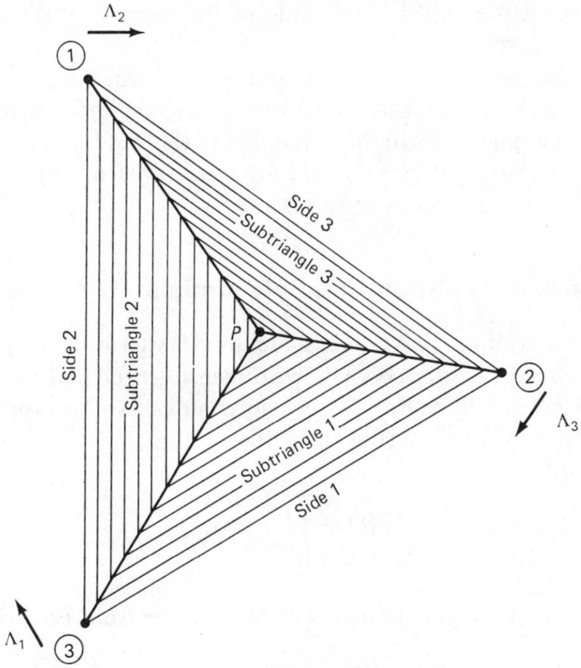

Fig. 4-5. Triangle for area coordinates.

Since the sum of the areas of the three subtriangles equals the area of the total triangle, we see that

$$\Lambda_1 + \Lambda_2 + \Lambda_3 = 1 \tag{4.4.3-7}$$

One can readily show that the area coordinates are identically equal to the linear shape functions; that is,

$$\Lambda_i(x, y) = p_i^1(x, y), \qquad 1 \leqslant i \leqslant 3 \tag{4.4.3-8}$$

From Equations (4.4.3-7) and (4.4.3-8), we see that

$$p_1^1(x, y) + p_2^1(x, y) + p_3^1(x, y) = 1 \tag{4.4.3-9}$$

Area coordinates and linear shape functions have the interesting property that

$$\Lambda_i(x, y) = p_i^1(x, y) = \text{constant}$$

along a line that is parallel to the side of the triangle along which these functions equal zero.

Since the area coordinates and the triangular linear shape functions are identical, they can be used interchangeably. It is sometimes convenient to do this. In certain applications in the remainder of this book (such as in Section 4.4.5), triangular linear shape functions are used in place of area coordinates (or, in this case, homogeneous coordinates).

4.4.4. Lagrange Interpolating Polynomials

Lagrange interpolating polynomials are used extensively in constructing higher order shape functions, both in two dimensions and in three dimensions. In general, the Lagrange interpolating polynomial can be expressed as the following continued product (Ref. 7, p. 201):

$$L_i^N(x) = \prod_{\substack{j=0 \\ j \neq i}}^{j=N} \frac{(x - x_j)}{(x_i - x_j)} \tag{4.4.4-1}$$

Here the degree of the polynomial is N. We can see from Equation (4.4.4-1) that

$$L_i^N(x_k) = \delta_{ik}, \qquad 0 \leqslant k \leqslant N \tag{4.4.4-2}$$

where, again, δ_{ik} is the Kronecker delta. From Equations (4.4.4-1) and (4.4.4-2), the polynomial $P(x)$, given by

$$P(x) = \sum_{i=0}^{i=N} f_i L_i^N(x) \tag{4.4.4-3}$$

interpolates exactly to the points (x_0, f_0), (x_1, f_1), (x_N, f_N) (see Figure 4-6).

In many of our higher order shape functions, we use polynomials that interpolate between uniformly spaced points. If the spacing between points is, say, the constant h, then

$$x_j = x_0 + jh \tag{4.4.4-4}$$

In this case, it becomes convenient to define the variable s by

$$x = x_0 + sh \tag{4.4.4-5}$$

We define the *normalized Lagrange polynomial*, $l_i^N(s)$ (Ref. 7, p. 201), by

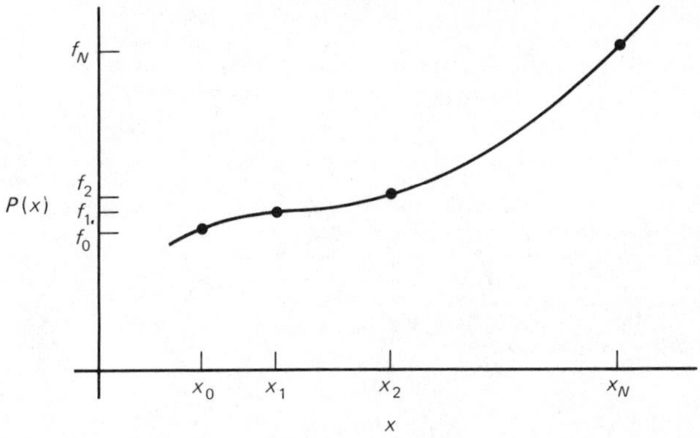

Fig. 4-6. Interpolating polynomial as a linear combination of Lagrange interpolating polynomials.

$$l_i^N(s) = L_i^N(x_0 + sh) = L_i^N(x) \tag{4.4.4-6}$$

From Equations (4.4.4-1) and (4.4.4-4) through (4.4.4-6), we see that

$$l_i^N(s) = \prod_{\substack{j=0 \\ j \neq i}}^{N} \frac{(s-j)}{(i-j)} \tag{4.4.4-7}$$

Finally, $l_N^N(s)$ is of particular interest with higher order shape functions over a triangle, as shown in the next section. From Equation (4.4.4-7), we have

$$l_N^N(s) = \prod_{j=0}^{N-1} \frac{(s-j)}{(N-j)} = \frac{1}{N!} \prod_{j=0}^{N-1} (s-j) \tag{4.4.4-8}$$

From Equation (4.4.3-8), we have, for the integer k, that

$$l_N^N(k) = 0, \qquad k < N \tag{4.4.4-9}$$

and that

$$l_N^N(N) = 1 \tag{4.4.4-10}$$

4.4.5. Higher Order Shape Functions over a Triangular Finite Element

There are various techniques reported in the literature (Ref. 2, pp. 65–67; Ref. 3, pp. 129–130) for the construction of higher order shape functions over a

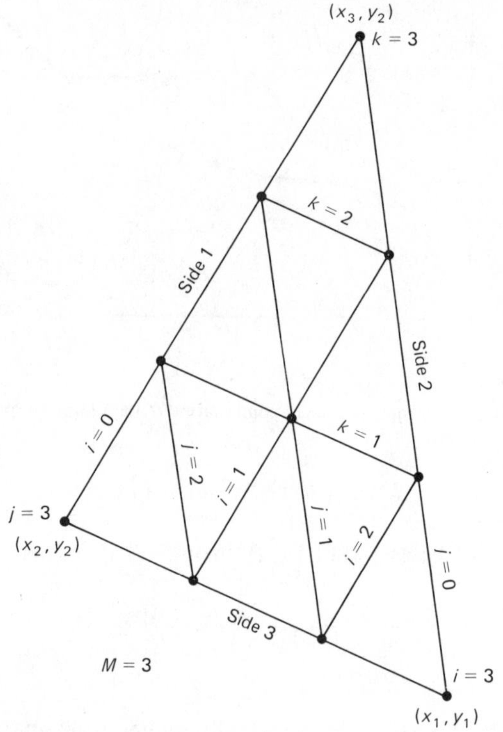

Fig. 4-7. Triangle for higher order shape function construction.

triangular finite element. While in appearance these techniques are different, the interpolating functions they produce are the same. The approach given here is essentially that of Silvester (Ref. 3, pp. 129–130).

The first step in constructing an Mth order shape function over a triangle is to create additional node points along the sides and perhaps in the interior of the triangle, as shown in Figure 4-7. (There are interior points if $M \geqslant 3$). To do this, we divide each side of the triangle into M equal intervals. We then join the points of subdivision of lines parallel to the sides of the triangle, as shown in the figure. This subdivides the triangle into M^2 congruent triangles. The vertices of these triangles are the $N_p = \frac{1}{2}(M + 1)(M + 2)$ node points of the triangle.

For identifying node points and developing the shape functions, it is convenient to number the lines in Figure 4-7. As shown, the integer i represents the number of a line parallel to side 1, j represents the number of a line parallel to side 2, and k represents the number of a line parallel to side 3. In this way, we can identify a node point as point (i, j, k), that is, the node point that is the intersection of lines i, j, and k. Furthermore, we have that

$$0 \leqslant i, j, k \leqslant M \tag{4.4.5-1}$$

In addition, for all points (x, y) along line i we have

$$i = Mp_1^1(x, y) \tag{4.4.5-2}$$

and similarly, along lines j and k,

$$j = Mp_2^1(x, y) \tag{4.4.5-3}$$

$$k = Mp_3^1(x, y) \tag{4.4.5-4}$$

And from these equations and Equation (4.4.3-9), we have

$$i + j + k = M \tag{4.4.5-5}$$

for any node point. From Equations (4.4.4-9), (4.4.4-10), and (4.4.5-2), we can see that $l_i^i[Mp_1^1(x, y)]$ (where l is a normalized Lagrange interpolating polynomial) equals unity along line i and equals zero along a line m that is parallel to line i, but for which $m < i$. Using Equations (4.4.5-3) and (4.4.5-4), we make similar statements about $l[Mp_2^1(x, y)]$ and $l[Mp_3^1(x, y)]$.

The shape function $\tau_{ijk}^{(M)}(x, y)$ is given by

$$\tau_{ijk}^{(M)}(x, y) = l_i^i[Mp_1^1(x, y)]l_j^j[Mp_2^1(x, y)]l_k^k[Mp_3^1(x, y)] \tag{4.4.5-6}$$

Since node point (i, j, k) is the intersection of lines i, j, and k, we see from the foregoing that if (x, y) is that node point, then

$$\tau_{ijk}^{(M)}(x, y) = l_i^i[Mp_1^1(x, y)] = l_j^j[Mp_2^1(x, y)] = l_k^k[Mp_3^1(x, y)] = 1$$

If (x, y) is any other node point, say, (m, n, o), we have, using Equation (4.4.5-5), that either $m < i$ and $l_i^i[Mp_1^1(x, y)] = 0$ or $n < j$ and $l_j^j[Mp_2^1(x, y)] = 0$ or $o < k$ and $l_k^k[Mp_3^1(x, y)] = 0$. In any case, from Equation (4.4.5-6),

$$\tau_{ijk}^{(M)}(x, y) = 0 \tag{4.4.5-7}$$

4.4.6. Shape Functions over a Rectangular Finite Element

These functions can easily be developed using Lagrange interpolating polynomials. For the rectangle with node points only at its vertices, we can see

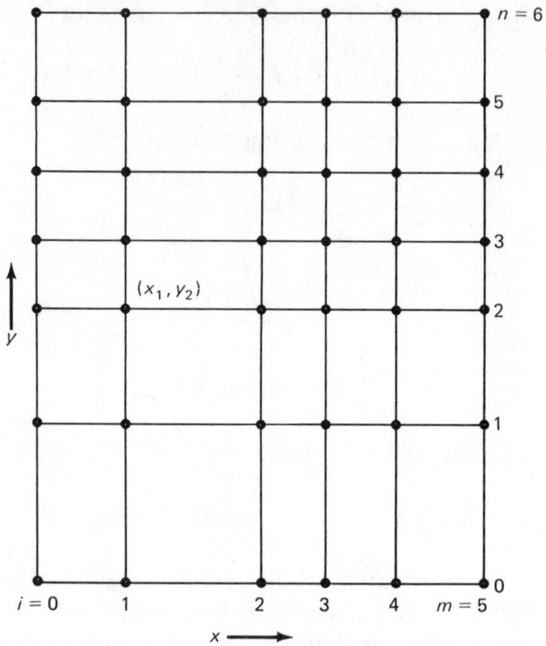

Fig. 4-8. Rectangle for higher order shape function construction.

that a set of its four shape functions are:

$$p_1(x, y) = L_0^1(x)L_0^1(y)$$

$$p_2(x, y) = L_1^1(x)L_0^1(y)$$

$$p_3(x, y) = L_1^1(x)L_1^1(y)$$

$$p_4(x, y) = L_0^1(x)L_1^1(y) \tag{4.4.6-1}$$

where L_0^1 and L_1^1 are Lagrange interpolating functions [defined in Equation (4.4.4-1)], the function over x interpolates from x_0 to x_1 and the function over y interpolates from y_0 to y_1.

The approach used to find lowest order shape functions for the rectangle is simply extended to find higher order shape functions for the rectangle in Figure 4-8. Here, we have $n + 1$ rows of points, each row having $m + 1$ points.

The coordinates of each point are (x_i, y_j) for $0 \leqslant i \leqslant m$ and $0 \leqslant j \leqslant n$. Again,

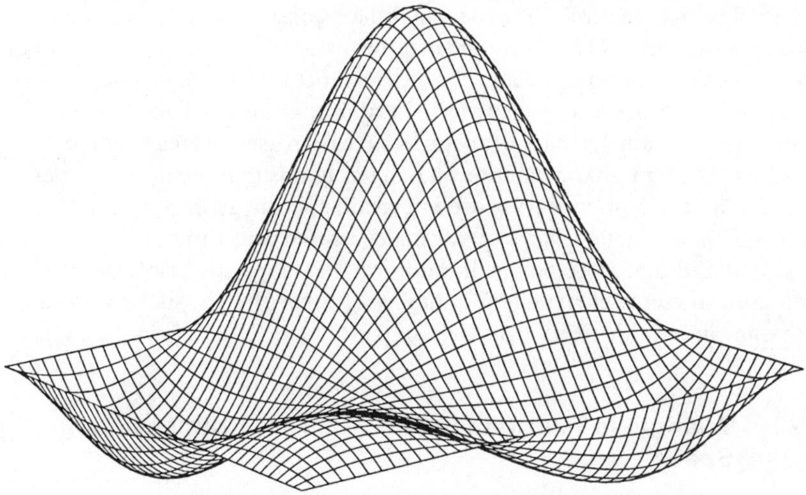

Fig. 4-9. Rectangular higher order shape function.

we use Lagrange interpolating polynomials. Then we have $(n + 1)(m + 1)$ shape functions given by

$$p_{ij}(x, y) = L_i^m(x)L_j^n(y) \tag{4.4.6-2}$$

In Equation (4.4.6-2), we have Lagrange interpolating polynomials, as discussed in Section 4.4.4. From this equation and Equation (4.4.4-2), we have

$$p_{ij}(x_k, y_l) = \delta_{ik}\delta_{jl} \tag{4.4.6-3}$$

where x_k and y_l are coordinates of interpolating points.

Figure 4-9 shows a typical rectangular higher order shape function.

4.5. ISOPARAMETRIC SHAPE FUNCTIONS IN TWO DIMENSIONS

All of the shape functions that have been formulated so far in this chapter have been functions of the spatial coordinates of the problem, the global coordinate system. As shown below, one can also use local, dimensionless *isoparametric* coordinates. When this is done, we express *both* the shape functions *and* the global coordinates in terms of these isoparametric coordinates. The disadvantage of using isoparametric coordinates is, of course, its complexity, compared to shape functions expressed directly in terms of the global coordinates. The advantage is in terms of the much greater flexibility one can achieve by being able to choose the functions by which one expresses the global coordinates in terms of the local, isoparametric coordinates.

Special isoparametric shape functions have been used to create, in the global coordinate system, shape functions having domains that are a wide variety of shapes. These are intended for use in certain specific geometrical configurations. Of these, triangles and quadrangles in two and three dimensions are presented and discussed below. In addition, the isoparametric approach has been used to create elements of a number of other shapes, such as a triangle with one curved side (Ref. 2, pp. 83–102).

4.5.1. Triangle in Two-dimensional or Three-dimensional Space

In two dimensions, we use the dimensionless, local, isoparametric coordinates ε and η, and, for convenience, we use the *standard triangle* defined in terms of these coordinates, shown in Figure 4-10. For the linear shape function $q_i^1(\varepsilon, \eta)$ over the standard triangle, we have

$$q_i^1(\varepsilon, \eta) = a_{1i} + a_{2i}\varepsilon + a_{3i}\eta, \qquad 1 \leqslant i \leqslant 3 \tag{4.5.1-1}$$

Furthermore,

$$q_i^1(\varepsilon_j, \eta_j) = \delta_{ij}, \qquad 1 \leqslant i, j \leqslant 3 \tag{4.5.1-2}$$

and, for Figure (4-10),

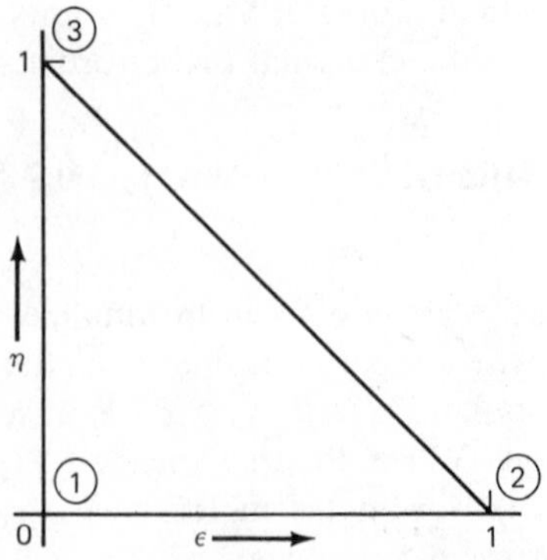

Fig. 4-10. Standard isoparametric triangle.

$$(\varepsilon_1, \eta_1) = (0, 0)$$

$$(\varepsilon_2, \eta_2) = (1, 0)$$

$$(\varepsilon_3, \eta_3) = (0, 1) \tag{4.5.1-3}$$

Combining Equations (4.5.1-1), (4.5.1-2), and (4.5.1-3) and solving the third-order linear systems yields

$$q_1^1(\varepsilon, \eta) = 1 - \varepsilon - \eta$$

$$q_2^1(\varepsilon, \eta) = \varepsilon$$

$$q_3^1(\varepsilon, \eta) = \eta \tag{4.5.1-4}$$

To obtain the shape function in global coordinates for the triangle in Figure 4-2, we use

$$p_i^1(x, y) = q_i^1(\varepsilon, \eta), \qquad 1 \leqslant i \leqslant 3 \tag{4.5.1-5}$$

$$x = q_1^1(\varepsilon, \eta)x_1 + q_2^1(\varepsilon, \eta)x_2 + q_3^1(\varepsilon, \eta)x_3 \tag{4.5.1-6}$$

$$y = q_1^1(\varepsilon, \eta)y_1 + q_2^1(\varepsilon, \eta)y_2 + q_3^1(\varepsilon, \eta)y_3 \tag{4.5.1-7}$$

Notice from Equations (4.5.1-5), (4.5.1-6), and (4.5.1-7) that (ε_1, η_1), (ε_2, η_2), and (ε_3, η_3), map in to (x_1, y_1), (x_2, y_2), and (x_3, y_3) identically.

We can map the standard triangle into a triangular plane segment in *three-dimensional space* with vertices (x_1, y_1, z_1), (x_2, y_2, z_2), and (x_3, y_3, z_3) simply by using Equations (4.5.1-6) and (4.5.1-7), in addition to the following equation (Ref. 8, p. 1462):

$$z = q_1^1(\varepsilon, \eta)z_1 + q_2^1(\varepsilon, \eta)z_2 + q_3^1(\varepsilon, \eta)z_3 \tag{4.5.1-8}$$

By replacing each shape function in Equations (4.5.1-6), (4.5.1-7), and (4.5.1-8) by an appropriate linear combination of shape functions of different orders, one can map into a triangle in three-dimensional space, with any desired curvature.

4.5.2. Quadrangle in Two- or Three-dimensional Space

We define a standard *rectangle* as shown in Figure 4-11, and again have the nondimensional coordinates ε and η. We have points (ε_1, η_1), (ε_2, η_2), (ε_3, η_3),

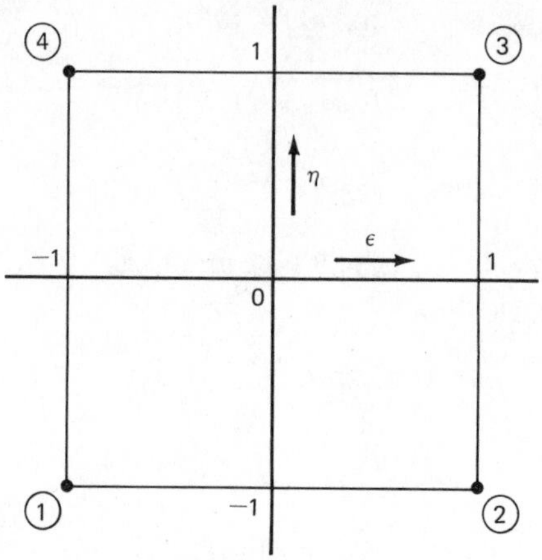

Fig. 4-11. Standard isoparametric rectangle.

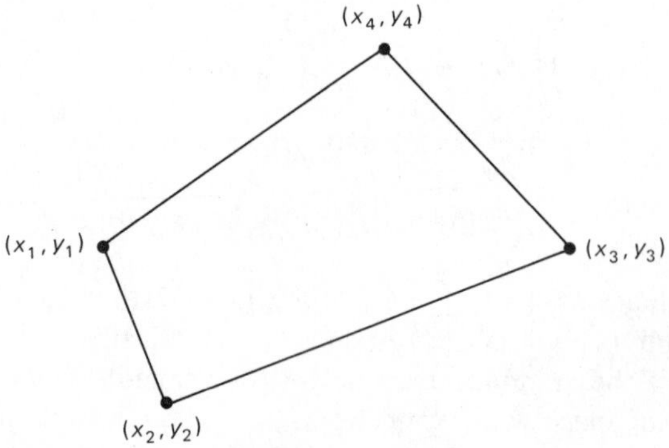

Fig. 4-12. Quadrilateral plane segment.

and (ε_4, η_4), as shown in this figure. We have the shape functions $q_i^1(\varepsilon, \eta)$, given by

$$q_1^1(\varepsilon, \eta) = \tfrac{1}{4}(1 - \varepsilon)(1 - \eta)$$

$$q_2^1(\varepsilon, \eta) = \tfrac{1}{4}(1 + \varepsilon)(1 - \eta)$$

$$q_3^1(\varepsilon, \eta) = \tfrac{1}{4}(1 + \varepsilon)(1 + \eta)$$

$$q_4^1(\varepsilon, \eta) = \tfrac{1}{4}(1 - \varepsilon)(1 + \eta) \qquad (4.5.2\text{-}1)$$

For the quadrilateral plane segment in Figure 4-12, we have

$$p_i^1(x, y) = q_i^1(\varepsilon, \eta), \qquad 1 \leqslant i \leqslant 4 \tag{4.5.2-2}$$

$$x = \sum_{i=1}^{4} x_i q_i^1(\varepsilon, \eta) \tag{4.5.2-3}$$

$$y = \sum_{i=1}^{4} y_i q_i^1(\varepsilon, \eta) \tag{4.5.2-4}$$

As with the triangle, we can map the standard rectangle into a quadrilateral plane segment in three-dimensional space simply by using the equations

$$p_i^1(x, y, z) = q_i^1(\varepsilon, \eta), \qquad 1 \leqslant i \leqslant 4 \tag{4.5.2-5}$$

and by using Equations (4.5.2-3) and (4.5.2-4) along with

$$z = \sum_{i=1}^{4} z_i q_i^1(\varepsilon, \eta) \tag{4.5.2-6}$$

And, as with the triangle, by replacing each shape function in Equations (4.5.2-3), (4.5.2-4), and (4.5.2-6) by an appropriate linear combination of shape functions of different orders, one can map the standard rectangle into a quadrilateral in three-dimensional space, of any desired curvature.

4.6. FINITE ELEMENTS AND SHAPE FUNCTIONS OF GLOBAL COORDINATES IN THREE-DIMENSIONAL PROBLEM DOMAINS

The two most popular finite elements in three dimensions are the tetrahedron and the rectangular parallelopiped (box). Accordingly, shape functions for these finite elements are given below. Discussions of interpolating functions for three-dimensional finite elements of other shapes can be found in books on the finite element method (Ref. 2, p. 108–110).

4.6.1. Subdivision of Problem Domain

Subdivision into rectangular parallelopipeds has the advantage of simplicity if the problem domain is shaped so that it can be covered by a small number of these finite elements. Subdivision into tetrahedra has the advantage that it will fit any irregularly-shaped problem domain. Furthermore, one can cover the three-dimensional problem domain by any irregular distribution of node points, and then connect these node points with straight-line segments in such a way as to form tetrahedra.

4.6.2. The Tetrahedron

The linear shape function over the tetrahedron is derived in much the same way as for the triangle. We give the tetrahedron vertices the indices 1, 2, 3, and 4. Then the shape functions p_j^1 are given by

$$p_j^1(x_i, y_i, z_i) = \delta_{ij}, \qquad 1 \leqslant i, j \leqslant 4 \tag{4.6.2-1}$$

where x_i, y_i, and z_i are coordinates of the vertex i. Since these shape functions are linear, they are given by

$$p_j^1(x, y, z) = c_{ij} + c_{2j}x + c_{3j}y + c_{4j}z, \qquad i \leqslant j \leqslant 4 \tag{4.6.2-2}$$

so that with Equation (4.6.2-1) we have

$$c_{ij} + c_{2j}x_i + c_{3j}y_i + c_{4j}z_i = \delta_{ij}, \qquad 1 \leqslant i, j \leqslant 4 \tag{4.6.2-3}$$

There are sixteen equations that comprise Equation (4.6.2-3), that is, four systems of four equations each. These systems are solved for the coefficients c_{1j}, c_{2j}, c_{3j}, and c_{4j}, $1 \leqslant j \leqslant 4$, by Cramer's rule (Ref. 4, p. 84), to be

$$p_j^1(x, y, z) = \frac{1}{\det M}[C_j + X_j x + Y_j y + Z_j z] \tag{4.6.2-4}$$

In Equation (4.6.2-4), M is a matrix given by

$$M = \begin{pmatrix} 1 & x_1 & y_1 & z_1 \\ 1 & x_2 & y_2 & z_2 \\ 1 & x_3 & y_3 & z_3 \\ 1 & x_4 & y_4 & z_4 \end{pmatrix} \tag{4.6.2-5}$$

where (x_1, y_1, z_1), (x_2, y_2, z_2), (x_3, y_3, z_3), and (x_4, y_4, z_4) are the coordinates of the vertices. The variables X_j, Y_j, and Z_j are the cofactors of x_j, y_j, and z_j in the matrix M_j and C_j is the cofactor of the 1 in row j (Ref. 4, p. 80).

Higher order shape functions for the tetrahedron have also been published (Ref. 2, p. 87).

4.6.3. The Rectangular Parallelopiped (Box)

The construction of shape functions over a box is very similar to the construction of shape functions over a rectangle (Section 4.5.3). We simply extend these

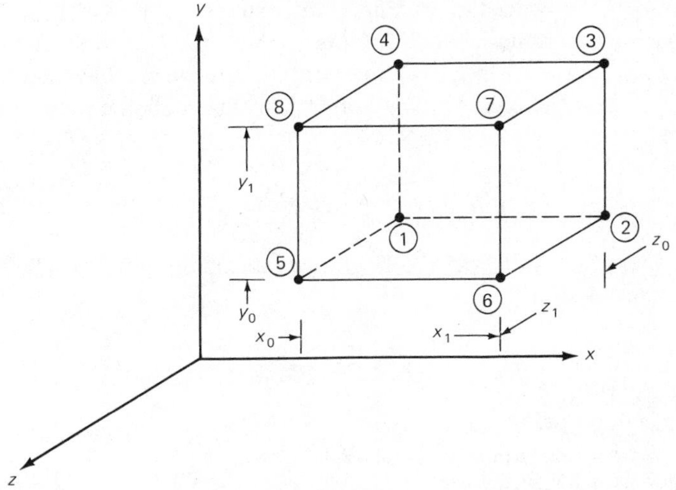

Fig. 4-13. Rectangular parallelopiped (box).

functions from two dimensions to three dimensions. The vertices of the box are indexed as shown in Figure 4-13. The lowest order shape functions (with node points only at the eight corners of the box) are:

$$p_1(x, y, z) = L_0^1(x)L_0^1(y)L_0^1(z) \tag{4.6.3-1}$$

$$p_2(x, y, z) = L_1^1(x)L_0^1(y)L_0^1(z) \tag{4.6.3-2}$$

$$p_3(x, y, z) = L_1^1(x)L_1^1(y)L_0^1(z) \tag{4.6.3-3}$$

$$p_4(x, y, z) = L_0^1(x)L_1^1(y)L_0^1(z) \tag{4.6.3-4}$$

$$p_5(x, y, z) = L_0^1(x)L_0^1(y)L_1^1(z) \tag{4.6.3-5}$$

$$p_6(x, y, z) = L_1^1(x)L_0^1(y)L_1^1(z) \tag{4.6.3-6}$$

$$p_7(x, y, z) = L_1^1(x)L_1^1(y)L_1^1(z) \tag{4.6.3-7}$$

$$p_8(x, y, z) = L_0^1(x)L_1^1(y)L_1^1(z) \tag{4.6.3-8}$$

where L_0 and L_1 are again Lagrange interpolating functions given by Equation (4.4.4-1).

The construction of higher order shape functions over a box is, again, an extension of the construction of higher order shape functions for a rectangle. We use a rectangular mesh of node points in a box, with x, y, and z coordinates $(x_1, x_2, \ldots, x_m)$, $(y_1, y_2, \ldots, y_n)$, $(z_1, z_2, \ldots, z_o)$ (where we can have nonuniform spacings between the points in x, y, and z). For this mesh, the shape functions are given by

$$p_{ijk}^{mno}(x, y, z) = L_i^m(x) L_j^n(y) L_k^o(z) \tag{4.6.3-9}$$

where $L_i^m(x)$, $L_j^n(y)$, and $L_k^o(z)$ are Lagrange interpolating polynomials, given by Equation (4.4.4-1).

REFERENCES

1. Steele, Charles W., and Mallinson, J. C., Theory of Low Permeability Heads, IEEE Transactions on Magnetics, September 1972, Vol. MAG-8, No.3.
2. Mitchell, A. R., and Wait, R. *The Finite Element Method in Partial Differential Equations.* New York: John Wiley & Sons, 1977.
3. Chari, M. V. K., and Silvester, P. P., Editors. *Finite Elements in Electrical and Magnetic Field Problems.* New York: John Wiley & Sons, 1980.
4. Nering, E. D. *Linear Algebra and Matrix Theory.* New York: John Wiley & Sons, 1965.
5. Zienkiewitz, O. C. *The Finite Element Method in Engineering Science.* New York: McGraw-Hill Book Company, 1971.
6. Vichevetsky, Robert. *Computer Methods for Partial Differential Equations.* Englewood Cliffs, N.J.: Prentice Hall, Inc., 1981.
7. Henrici, P. *Introduction to Numerical Analysis.* New York: John Wiley & Sons, 1964.
8. Jeng, G., and Wexler, A. *Isoparametric, Finite Element Variational Solution of Integral Equations for Three-Dimensional Fields*, International Journal for Numerical Methods in Engineering, Vol. 11, 1977, pp. 1455–1471.

5
PROJECTION METHODS IN FIELD COMPUTATIONS

5.1. INTRODUCTION

The objective of this chapter is to demonstrate, in a very basic way, how the finite element method and the integral equation method work to provide us with numerical solutions to linear field problems. These demonstrations are made in a very general way by the use of operators that represent the operations of differentiation and integration that are encountered in the equations that we solve.

These demonstrations show the following:

1. That each of our numerical solutions to a linear field problem, whether by the finite element method or the integral equation method, is a linear projection of the exact solution onto a certain finite dimensional linear space.
2. That it is often useful to define a norm of the error. This error is the difference between the exact solution and the numerical solution.
3. That certain numerical solutions (of both finite element problems and integral equation problems) are *orthogonal* projections. Furthermore, an orthogonal projection is that numerical solution that minimizes a certain error norm.

Sections 5.2 and 5.3 provide certain commonly known mathematical concepts that are necessary to understand the algorithms that we develop for field computation. Section 5.2 presents certain special linear spaces that are used in developing the finite element method, the integral equation method, and norms of the error in their solutions. Section 5.3 gives a definition of the concept of an operator. As shown in this chapter, the starting equation for either the finite element method or the integral equation method can be represented by a simple operator equation. In this operator equation, there is one operator that represents either the operations of differentiation (in the finite element method) or the operations of integration (in the integral equation method). These operator equations provide a concise, straightforward means for uniform presentation and discussion of the basic algorithms that we use in field computations.

Section 5.4 presents certain approaches that are used, in conjunction with the finite element method and the integral equation method, to develop our algorithms. These approaches are the Rayleigh-Ritz approach and the Galerkin and related approaches. The Galerkin approach is used throughout the remainder of this book, with the thought that it is simpler and more flexible than the Rayleigh-Ritz approach.

Sections 5.5 and 5.6 present the finite element method and the integral equation method, respectively. These methods are presented in a general way, with the use of operator equations.

The reader may wonder why these methods are presented in this chapter and again in Chapters 6, 7, and 8. The reason is that the objectives of the presentation are very different here from those sought in Chapters 6, 7, and 8. Here the objectives are as follows:

1. To present the basic elements of these methods.
2. To show certain similarities between the finite element method and the integral equation method.
3. To develop certain equations from which we can show that the numerical solutions to linear problems, achieved either by the finite element method or by the integral equation method, are, in fact, linear projections of the exact solution onto a finite dimensional subspace.

Alternatively, the objectives of Chapters 6, 7, and 8 are to present the finite element method and the integral equation method in sufficient detail to provide the reader with a practical working knowledge of these methods. The degree of detail presented in those chapters is necessarily so great that the *basic properties* of the methods can be easily overlooked by the reader. Putting it another way, in those chapters, the reader can easily fail to "see the forest because of the trees."

Because of the use of operator notation in Sections 5.5 and 5.6, the uninitiated reader may find the presentations of the finite element method and the integral equation method somewhat vague and abstract. If so, the reader is encouraged to turn to Chapters 6 and 8, where many of the commonly used differential and integral operators are presented and defined.

The objective of Section 5.7 is to show that the numerical solutions to linear field problems (by either the finite element method or the integral equation method) are, in fact, linear projections of the exact solution onto a finite dimensional subspace. This is done by first defining a linear projection, and then showing that these numerical solutions to linear problems conform to that definition.

Section 5.8 first defines and discusses an *orthogonal* linear projection. It then shows that the error that results with an orthogonal projection has a smaller error norm than the error that results with any nonorthogonal projection. Finally, it uses the finite element method and the integral equation method to differentiate which of the solutions presented in this book are orthogonal projections and which are nonorthogonal.

5.2. SPECIAL LINEAR SPACES IN FIELD COMPUTATIONS

5.2.1. Metric Spaces

In a *metric* space, we define a *distance* $d(x, y)$ between any two elements x and y of the space (Ref. 1, pp. 99–101). Then $d(x, y)$ is a function of x and y that has the following properties:

$$d(x, y) = d(y, x) \tag{5.2.1-1}$$

$$d(x, y) \geqslant 0 \tag{5.2.1-2}$$

$$d(x, y) = 0 \text{ if and only if } x = y \tag{5.2.1-3}$$

$$d(x, z) \leqslant d(x, y) + d(y, z) \tag{5.2.1-4}$$

A metric space is not necessarily a linear space.

In field computations, the exact field, $\boldsymbol{\gamma}$, and the numerically computed approximate field, $\boldsymbol{\beta}$, are elements of the same linear space. If this linear space is a metric space, then we have defined the distance $d(\boldsymbol{\gamma}, \boldsymbol{\beta})$, between the exact field and the numerically computed field. As shown below, this distance is a convenient measure of the error in our calculation. In addition, this distance is quite significant for certain algorithms that we use.

5.2.2. Completeness of a Metric Space

Consider a sequence of element x_k in a metric space. We say that x_k *converges* to x if, for any $\varepsilon > 0$, there is an index, N, such that

$$d(x, x_k) < \varepsilon \qquad \text{for} \qquad k > N$$

Furthermore, a sequence $\{x_k\}$ is a *Cauchy* sequence if, for every $\varepsilon > 0$, there is an index, N, for which

$$d(x_m, x_p) \leqslant \varepsilon \qquad \text{for} \qquad m, p > N$$

We can show that if a sequence $\{x_k\}$ converges, it is a Cauchy sequence (Ref. 1, p. 100).

We say that a metric space is *complete* if every Cauchy sequence in that space converges to an element in that space.

To provide a feeling for the meaning of completeness, Stakgold (Ref. 1, pp. 101, 102) exhibits a metric space that is not complete. Specifically, it is the space of rational numbers having a metric

$$d(x, y) = |x - y|$$

He demonstrates that this space is incomplete by finding a Cauchy sequence within the space that converges to an irrational number. One can think of the space of rational numbers as being imcomplete because it is full of infinitesimal gaps that would otherwise be occupied by the irrational numbers.

5.2.3. Normed Linear Spaces

A normed linear space is a linear space in which we define the real-valued function $\|x\|$, known as the *norm* of x. It has the properties

$$\|x\| > 0 \tag{5.2.3-1}$$

$$\|x\| = 0 \qquad \text{if and only if } x = 0 \tag{5.2.3-2}$$

$$\|bx\| = |b|\,\|x\| \tag{5.2.3-3}$$

for any real number b, and

$$\|x_1 + x_2\| \leqslant \|x_1\| + \|x_2\| \tag{5.2.3-4}$$

Notice that a normed linear space is a metric space if we defined the distance $d(x, y)$ by

$$d(x, y) = \|x - y\| \tag{5.2.3-5}$$

We call $d(x, y)$ as given by Equation (5.2.3-5) the *natural metric generated by the norm.*

The definition of distance given by Equation (5.2.3-5) is a convenient one to use in the study of our algorithms. Suppose that $\boldsymbol{\gamma}$ is the exact solution to our problem and that $\boldsymbol{\beta}$ is the numerically computed solution. Then, as we see below, a convenient measure of the error in our computation is

$$\|\boldsymbol{\gamma} - \boldsymbol{\beta}\| = d(\boldsymbol{\gamma}, \boldsymbol{\beta}) \tag{5.2.3-6}$$

We call ε, given by

$$\varepsilon = \gamma - \beta$$

the *error* and the norm of ε, as given in Equation (5.2.3-6), the *error norm.*

5.2.4. Inner Product Spaces, Hilbert Spaces, and Orthogonality

An *inner product* $\langle x, y \rangle$ on a *real* linear space is a real-valued, symmetric, positive-definite bilinear form. That is, it has the following properties:

$$\langle x, y \rangle = \langle y, x \rangle \tag{5.2.4-1}$$

$$\langle bx, y \rangle = b\langle x, y \rangle \tag{5.2.4-2}$$

$$\langle x_1 + x_2, y \rangle = \langle x_1, y \rangle + \langle x_2, y \rangle \tag{5.2.4-3}$$

$$\langle x, x \rangle \geqslant 0 \tag{5.2.4-4}$$

$$\langle x, x \rangle = 0, \qquad \text{if and only if } x = 0 \tag{5.2.4-5}$$

A linear space for which we define an inner product is an *inner product space.* From Equations (5.2.3-1), (5.2.3-2), (5.2.3-3), and (5.2.3-4) and Equations (5.2.4-2), (5.2.4-3), (5.2.4-4), and (5.2.4-5), as well as the Schwartz inequality,* we can show that

$$\langle x, x \rangle^{1/2}$$

is a norm on the inner product space. That is, the inner product has a *natural norm,*

$$\|x\| = \langle x, x \rangle^{1/2} \tag{5.2.4-6}$$

And since a norm generates a natural metric, we have

$$d(x, y) = \|x - y\| = \langle x - y, x - y \rangle^{1/2} \tag{5.2.4-7}$$

---

*The Schwartz inequality states that in any inner product space, for x and y in that space $|\langle x, y \rangle|^2 \leqslant \langle x, x \rangle \langle y, y \rangle$

An inner product space that is complete in its natural metric is called a *Hilbert space.*

The elements x and y of an inner product space (or Hilbert space) are *orthogonal* to each other if

$$\langle x, y \rangle = 0 \tag{5.2.4-8}$$

As shown below, this property of orthogonality plays an important part in many of the algorithms that we use in field computations.

5.3. OPERATORS IN FIELD CALCULATIONS

The value of operators is that they are a convenient way to represent field problems (and other problems as well) in a simple, unified way. That is, a field problem can be represented by the equation

$$K\gamma = \mathbf{g} \tag{5.3-1}$$

where γ is the exact solution (the field) and K is an operator. If K is a differential operator, this equation represents a differential equation, and if K is an integral operator, this equation represents an integral equation. An objective of this section is then to discuss operators and their use in representing integral and differential equations, that is, integral equation and finite element field problems. Each field problem, whether it is solved by the finite element method or by the integral equation method, will be represented by an equation of the form of Equation (5.3-1). A further objective is that this equation will be suitable for use in developing the projection method.

Basically an operator is a certain type of function. For example, a function, f, in this equation

$$y = f(x)$$

maps elements x of the set X into the element y in the set Y. That is, to each element x in the set X, we associate our element y in the set Y. The set X is called the *domain*, D, of the function f, and the set Y is called the *range*, R, of function f.

In the same way, an operator K maps an element γ of the set S into another element $\mathbf{g}$ of the set S. We write this as:

$$K\gamma = \mathbf{g} \tag{5.3-2}$$

Again, the set of such elements γ is called the *domain* of K, and the set of such elements $\mathbf{g}$ of S is called the *range* of K.

An operator is *one to one* if for any element $\mathbf{g}$ in R there corresponds one and only one γ in D. A *one-to-one* operator is therefore *invertible*. That is, one can define the inverse of K, K^{-1}, by

$$K^{-1}\mathbf{g} = \gamma \tag{5.3-3}$$

Then the range of K is the domain of K^{-1} and the domain of K is the range of K^{-1}.

An operator is *linear* if

$$K(\gamma_1 + \gamma_2) = K\gamma_1 + K\gamma_2 \tag{5.3-4}$$

and

$$K(b\gamma) = bK(\gamma) \tag{5.3-5}$$

where b is a scalar. Notice that, in order for Equations (5.3-4) and (5.3-5) to hold for all γ in D, it is necessary for D and R to be *linear spaces*. Thus we say that an operator is linear *only* if its range and domain are linear spaces. In other words, two requirements must be met in order for an operator to be linear:

1. The operation that it presents (differentiation, integration, etc.) must be linear.
2. The domain and the range of the operator must be linear spaces.

5.4. APPROACHES USED IN OBTAINING APPROXIMATE SOLUTIONS TO FIELD PROBLEMS

This section presents, in general terms, the approaches that are used in obtaining solutions to linear field problems. These approaches are used with both the finite element method and the integral equation method.

In general, the steps used in a numerical linear field problem (with either the finite element method or the integral equation method) are the following:

1. Define the problem (as discussed in Chapter 3) and decide which method of solution to use (the finite element method or the integral equation method). This includes defining the field to be computed, the equation

to be solved, the problem domain, and, if applicable, any boundary conditions.

2. Define a Hilbert space, $\mathscr{H}$, for the problem.* This means defining the elements of $\mathscr{H}$, and an inner product between any two elements. Each element of $\mathscr{H}$ is a function of spatial coordinates, defined over the problem domain. As shown below, these elements, when used with the finite element method, are required to satisfy certain boundary conditions.
3. Deploy node points and finite elements over the problem domain. Decide which of these node points are *active node points* (at which the field or source distribution is to be computed, as discussed below). (The number of active node points is N_a.)
4. Define shape functions and local-support basis functions (as discussed in Chapter 4) over these finite elements. Precisely one basis function corresponds to each node point.
5. Based upon the approach and method used, the equation to be solved, and the basis functions defined, construct a set of linear equations in terms of the field or source values at the active node points. The number of these equations equals the number of active node points.
6. Solve this set of linear equations for the fields or sources at the active node points.
7. If the integral equation method is used, then compute the field at any desired point, based upon the sources computed in step 6.

While certain of these steps may seem vague, abstract, and difficult to comprehend at this point, they will be greatly clarified as they are exemplified many times in the remainder of this book.

Some comments are in order concerning the Hilbert space discussed in step 2, above. Nearly every numerical field problem that is solved by any of the approaches discussed in this section, and by either the integral equation method or the finite element method, has associated with it a specific Hilbert space.* On the other hand, many papers on field computations do not mention the Hilbert spaces that are associated with their algorithms, because there is no need to do so. Furthermore, the reader may well feel "Why define a Hilbert space when what I *really* want to do is compute a field?" Certainly, if the reader can complete steps 1, 3, 4, and 5 above, without difficulty, then there is no need to go through the formality of defining the elements and inner product of the Hilbert space. Suppose, however, that, for some problem, the reader has difficulty in understanding how to complete these steps. It is the author's

*In the collocation approach, as described in Section 5.6.2 below, we do not define an inner product. Therefore, with this approach and the integral equation method, we define a *linear* space, but not a *Hilbert* space.

belief that, in this circumstance, the careful definition of the Hilbert space, that is, its elements and its inner product, helps substantially to clarify the construction of the algorithm and the solution of the problem. It is for this reason that Hilbert spaces are defined for all algorithms and problems presented in the remainder of this book.

Referring to step 3 above, all of the node points used in the integral equation method are active node points. In the finite element method, the field values at certain node points are set at the outset of the problem, in order to satisfy certain Dirichlet boundary conditions. This is done in the algorithms of Chapters 6, 7, and 10, where these node points are called *Dirichlet node points.* All remaining node points (a majority) in a finite element problem are called active node points. With either the integral equation method or the finite element method, the field or source density at all active node points is computed in step 6 above.

In step 4, we construct a basis function for each node point. These basis functions form the basis of a finite dimensional Hilbert space, $\mathscr{H}_N$, which is a subspace of $\mathscr{H}$. The dimension of $\mathscr{H}_N$ is just equal to the number of node points in the problem.

There are two approaches commonly used in step 5 to construct the set of linear equations for a field computation. The first, the Rayleigh-Ritz approach, is presented in Section 5.4.1, whereas the Galerkin and associated approaches are presented in Section 5.4.2. For a given problem, as defined in steps 1, 2, 3, and 4 above, the Rayleigh-Ritz approach and the Galerkin approach lead to the same set of linear equations in step 5. After our discussion here, we use the Galerkin approach throughout the remainder of this book. We feel the Galerkin approach has two advantages over the Rayleigh-Ritz:

1. It is the simpler approach to implement.
2. It can be applied readily to problems with operators that are not self-adjoint.

It can be shown that, for most finite element field problems, including those discussed in this book, operators used with inhomogenous (nonzero) Dirichlet boundary conditions are not self-adjoint.

From this point to the end of this chapter, operator notation, as discussed in Section 5.3, is used. That is, operator notation is used to represent both the operation of differentiation for the finite element method and the operation of integration for the integral equation method. In fact, many of the operators that appear in the remainder of this chapter can represent *either* integrations or differentiations. Likewise, many of the operator equations that appear here can represent either the finite element method or the integral equation method. In this way, the use of operator notation is a great help in providing us with a unified view of numerical field computation.

5.4.1. Rayleigh-Ritz Approach

This section presents and discusses the Rayleigh-Ritz method of solution, as it is applied to integral equation problems. This method has been applied to field computations by both the integral equation method and the finite element method (Refs. 5, 6). (The application of the Rayleigh-Ritz approach to the finite element method differs somewhat from the derivation given here, due to its use of an integration by parts.)

Suppose that in the equation

$$K\boldsymbol{\gamma} = \mathbf{g} \tag{5.4.1-1}$$

$\boldsymbol{\gamma}$ and $\mathbf{g}$ are elements of a Hilbert space, $\mathscr{H}$, and that the domain of the operator K, D_K, is precisely $\mathscr{H}$. Suppose furthermore that K is linear, self-adjoint, and positive definite. By self-adjoint, we mean that, for any elements $\mathbf{u}$ and $\mathbf{v}$ of $\mathscr{H}$,

$$\langle K\mathbf{u}, \mathbf{v}\rangle = \langle \mathbf{u}, K\mathbf{v}\rangle \tag{5.4.1-2}$$

By positive definite, we mean that for any element $\mathbf{u}$ of $\mathscr{H}$,

$$\langle K\mathbf{u}, \mathbf{u}\rangle \geqslant 0 \tag{5.4.1-3}$$

Since K is positive definite, we know that Equation (5.4.1-1) has only one solution (Ref. 6, p. 74).

Consider the quadratic functional, given by

$$F(\mathbf{u}) = \langle K\mathbf{u}, \mathbf{u}\rangle - 2\langle \mathbf{g}, \mathbf{u}\rangle \tag{5.4.1-4}$$

In Equation (5.4.1-4), $\mathbf{u}$ is, again, an element of $\mathscr{H}$; and F is called a functional because its domain D_F lies in a linear space (in this case, $\mathscr{H}$) and its range lies in the space of real numbers. By the *minimal functional theorem* (Ref. 8, pp. 74–76, and Ref. 1, pp. 245–246), we have two results:

1. If $F(\mathbf{u})$ attains its minimum value for some function $\boldsymbol{\gamma}$, then $\boldsymbol{\gamma}$ is a solution of Equation (5.4.1-1).
2. If Equation (5.4.1-1) has a solution, $\boldsymbol{\gamma}$, then $F(\mathbf{u})$ attains its minimum value for $\mathbf{u} = \boldsymbol{\gamma}$.

These results are easy to prove. For result 1, suppose that

$$F(\boldsymbol{\gamma}) \leqslant F(\mathbf{u}) \tag{5.4.1-5}$$

where **u** is any element of $\mathscr{H}$. Then we can represent **u** as

$$\mathbf{u} = \gamma + \varepsilon\boldsymbol{\eta} \tag{5.4.1-6}$$

where ε is a real number and $\boldsymbol{\eta}$ is any element of $\mathscr{H}$, and from this equation,

$$\frac{d\mathbf{u}}{d\varepsilon} = \boldsymbol{\eta} \tag{5.4.1-7}$$

From Equations (5.4.1-4) and (5.4.1-7), we have

$$\frac{dF(\mathbf{u})}{d\varepsilon} = \langle K\mathbf{u}, \boldsymbol{\eta}\rangle + \langle K\boldsymbol{\eta}, \mathbf{u}\rangle - 2\langle \mathbf{g}, \boldsymbol{\eta}\rangle$$

and since K is self-adjoint, we have

$$\frac{dF(\mathbf{u})}{d\varepsilon} = 2\langle K\mathbf{u} - \mathbf{g}, \boldsymbol{\eta}\rangle$$

and with Equation (5.4.1-6),

$$\frac{dF(\mathbf{u})}{d\varepsilon} = 2\langle k\gamma - \mathbf{g}, \boldsymbol{\eta}\rangle + 2\varepsilon\langle K\boldsymbol{\eta}, \boldsymbol{\eta}\rangle \tag{5.4.1-8}$$

Since γ minimizes the functional, we see, from Equation (5.4.1-6), that

$$\left.\frac{dF(\mathbf{u})}{d\varepsilon}\right|_{\varepsilon=0} = 0$$

and this, with Equation (5.4.1-8), yields

$$\langle k\gamma - \mathbf{g}, \boldsymbol{\eta}\rangle = 0 \tag{5.4.1-9}$$

Since $\boldsymbol{\eta}$ can be any element of $\mathscr{H}$, we see that

$$K\gamma - \mathbf{g} = 0 \tag{5.4.1-10}$$

which was to be proven.

For result 2, we assume that

$$K\gamma = \mathbf{g} \tag{5.4.1-11}$$

Substituting **g** from Equation (5.4.1-11) into Equation (5.4.1-4), we have

$$F(\mathbf{u}) = \langle K\mathbf{u}, \mathbf{u} \rangle - 2\langle K\boldsymbol{\gamma}, \mathbf{u} \rangle$$

After several steps of manipulation that use the self-adjoint property of K, we obtain

$$F(\mathbf{u}) = \langle K(\mathbf{u} - \boldsymbol{\gamma}), \mathbf{u} - \boldsymbol{\gamma} \rangle - \langle K\boldsymbol{\gamma}, \boldsymbol{\gamma} \rangle$$

From this equation, letting $\mathbf{u} = \boldsymbol{\gamma}$, we have

$$F(\boldsymbol{\gamma}) = -\langle K\boldsymbol{\gamma}, \boldsymbol{\gamma} \rangle$$

so that

$$F(\mathbf{u}) = \langle K(\mathbf{u} - \boldsymbol{\gamma}), \mathbf{u} - \boldsymbol{\gamma} \rangle + F(\boldsymbol{\gamma})$$

Finally, from this equation and the positive definiteness of K, we have that

$$F(\mathbf{u}) \geqslant F(\boldsymbol{\gamma})$$

which was to be proven.

One can use the first of these results to obtain the exact solution to Equation (5.4.1-1). To do this, one would (a) construct the functional $F(\mathbf{u})$ in Equation (5.4.1-4) and (b) find the value of **u** that minimizes this functional. This is called the *variational* method of solution (Ref. 1, p. 245).

By a small step of logic, we can adapt the variational method for finding the exact solution, $\boldsymbol{\gamma}$, to a method for finding the approximate solution, $\boldsymbol{\beta}$. Here, $\boldsymbol{\beta}$ is the linear combination of N basis elements given by Equation (4.2-1). (That is, $\boldsymbol{\beta}$ is an element of $\mathscr{H}_N$, which is an N-dimensional subspace of the Hilbert space $\mathscr{H}$, where $\mathscr{H}_N$ has the same elements as the subspace S_N, that is, $\mathscr{H}_N$ has, as a basis, the basis functions $\boldsymbol{\alpha}_1, \boldsymbol{\alpha}_2, \ldots, \boldsymbol{\alpha}_N$.)

Specifically, we let any element **w** of $\mathscr{H}_N$ be given by

$$\mathbf{w} = \sum_{i=1}^{N} w_i \boldsymbol{\alpha}_i \qquad (5.4.1\text{-}12)$$

and then we find $\boldsymbol{\beta}$, where

$$\boldsymbol{\beta} = \sum_{i=1}^{N} \beta_i \boldsymbol{\alpha}_i \qquad (5.4.1\text{-}13)$$

so that

$$F(\boldsymbol{\beta}) \leqslant F(\mathbf{w}) \tag{5.4.1-14}$$

for all $\mathbf{w}$ in $\mathscr{H}_N$. This method for calculating $\boldsymbol{\beta}$ is known as the *Rayleigh-Ritz* method (Ref. 1, pp. 248–250). Our intuition tells us that since the exact solution can be obtained by minimizing $F(\mathbf{u})$ over $\mathscr{H}$, then a good approximate solution can be obtained by minimizing $F(\mathbf{w})$ over $\mathscr{H}_N$.

Since the operator K is positive definite, we can find the value of $\mathbf{w}$ that minimizes $F(\mathbf{w})$ simply by making $F(\mathbf{w})$ stationary with respect to the coefficient $w_1, w_2, \ldots, w_N$ in Equation (5.4.1-12). That is, we find the $\mathbf{w}$ for which

$$\frac{\partial F(\mathbf{w})}{\partial w_i} = 0, \qquad 1 \leqslant i \leqslant N \tag{5.4.1-15}$$

Combining Equations (5.4.1-4) and (5.4.1-15) yields

$$\frac{\partial F(\mathbf{w})}{\partial w_i} = \left\langle K\mathbf{w}, \frac{\partial \mathbf{w}}{\partial w_i} \right\rangle + \left\langle K\frac{\partial \mathbf{w}}{\partial w_i}, \mathbf{w} \right\rangle - 2\left\langle \mathbf{g}, \frac{\partial \mathbf{w}}{\partial w_i} \right\rangle = 0$$

and since K is self-adjoint, we have

$$0 = \left\langle K\mathbf{w} - \mathbf{g}, \frac{\partial \mathbf{w}}{\partial w_i} \right\rangle, \qquad 1 \leqslant i \leqslant N \tag{5.4.1-16}$$

Furthermore, differentiating Equation (5.4.1-12) yields

$$\frac{\partial \mathbf{w}}{\partial w_i} = \boldsymbol{\alpha}_i$$

so that Equation (5.4.1-16) becomes

$$0 = \langle K\mathbf{w} - \mathbf{g}, \boldsymbol{\alpha}_i \rangle, \qquad 1 \leqslant i \leqslant N \tag{5.4.1-17}$$

Since the function $\mathbf{w}$ in Equation (5.4.1-17) is just that element of $\mathscr{H}_N$ that minimizes $F(\mathbf{w})$, then this $\mathbf{w}$ equals $\boldsymbol{\beta}$ [from Inequality (5.4.1-14)], so that

$$0 = \langle K\boldsymbol{\beta} - \mathbf{g}, \boldsymbol{\alpha}_i \rangle, \qquad 1 \leqslant i \leqslant N \tag{5.4.1-18}$$

From this equation and Section 5.7, we see that the Rayleigh-Ritz method is, in fact, a projection method. In fact, these equations represent the *Galerkin method.* The system of Equations (5.4.1-18), when combined with Equation (4.2-1), yields a system of N linear equations that can then be solved for the coefficients $\beta_1, \beta_2, \ldots, \beta_N$. These coefficients then constitute the solution.

5.4.2. Galerkin and Related Approaches

The Galerkin and related approaches are used with both the finite element method and the integral equation method to produce the algorithms that we use. As shown here and in the next three sections, when these methods are applied to linear field problems, the resulting approximate solution, $\boldsymbol{\beta}$, is a linear projection of the exact solution, $\boldsymbol{\gamma}$, onto the finite-dimensional Hilbert space, $\mathscr{H}_N$.

When one of the Galerkin or related approaches is used in step 5 (presented at the beginning of Section 5.4), the subsequent steps are these:

a. Construct a system of equations in terms of the *exact* solution, $\boldsymbol{\gamma}$. The number, N_a, of these equations equals the number of active node points.
b. From this system, construct a system of N_a linear equations in terms of the approximate solution, $\boldsymbol{\beta}$, in such a way that $\boldsymbol{\beta}$ is a linear projection of $\boldsymbol{\gamma}$ onto $\mathscr{H}_N$.

As shown in Sections 5.5 and 5.6, the formulation by which we proceed from step a to step b is different for the finite element method from what it is for the integral equation method. For this reason, only the systems of equations of step a are presented here.

As before, the operator equation to be solved is

$$K\boldsymbol{\gamma} = \mathbf{g} \tag{5.4.2-1}$$

where K is an operator on $\mathscr{H}$, and $\boldsymbol{\gamma}$ and $\mathbf{g}$ are elements of $\mathscr{H}$.

For the Bubnov-Galerkin method (or Galerkin method), we have the system (Ref. 5, p. 367)

$$\langle K\boldsymbol{\gamma} - \mathbf{g}, \boldsymbol{\alpha}_i \rangle = 0, \qquad 1 \leqslant i \leqslant N_a \tag{5.4.2-2}$$

In this system, the $\boldsymbol{\alpha}_i$ for $1 \leqslant i \leqslant N_a$ are the basis functions that correspond to the active node points.

The method of moments is given by (see Ref. 5, p. 367, and Ref. 6, pp. 5–7)

$$\langle K\boldsymbol{\gamma} - \mathbf{g}, S\boldsymbol{\alpha}_i \rangle = 0, \qquad 1 \leqslant i \leqslant N_a \tag{5.4.2-3}$$

where S is an operator on $\mathscr{H}$. In the particular case in which $S = K$, we have the method of least squares (Ref. 5, p. 367):

$$\langle K\gamma - \mathbf{g}, K\boldsymbol{\alpha}_i \rangle = 0, \qquad 1 \leqslant i \leqslant N_a \tag{5.4.2-4}$$

In all of the approaches presented here in Section 5.4.2, the approximate solution $\boldsymbol{\beta}$, mentioned in step b above, is assumed to be a linear combination of the basis functions $\boldsymbol{\alpha}_i$, for $1 \leqslant i \leqslant N_a$. One of the advantages of the approaches presented above is that the elements of $\mathscr{H}$ that are used as the second arguments of the inner products are either $\boldsymbol{\alpha}_i$ or the result of some operation on $\boldsymbol{\alpha}_i$. This means that to construct these second arguments, we can use the same node points and finite elements that we used to construct the $\boldsymbol{\alpha}_i$'s. And, for the Galerkin method, we can use the $\boldsymbol{\alpha}_i$'s themselves as these second arguments.

Finally, we have the *method of weighted residuals*, given by Reference 5, (p. 367):

$$\langle K\gamma - \mathbf{g}, \mathbf{p}_i \rangle = 0, \qquad 1 \leqslant i \leqslant N_a \tag{5.4.2-5}$$

Here the $\mathbf{p_i}$ for $1 \leqslant i \leqslant N_a$ comprise another set of basis functions (that are not necessarily in $\mathscr{H}_N$). (In general, the supports for the $\mathbf{p_i}$ do not need to be the same as those for the $\boldsymbol{\alpha}_i$. And the structure of node points and finite elements used for $\mathbf{p_i}$ do not need to be the same as those used for $\boldsymbol{\alpha}_i$.) There is no advantage to constructing $\mathbf{p_i}$ as another set of pyramid-type basis functions—distinct from $\boldsymbol{\alpha}_i$. That would only increase the amount of computation required by the problem, with no advantage to the solution. The method of weighted residuals has only one common application. This is the *collocation approach*, which is used with the integral equation method. It is presented in Section 5.6.

5.5. FINITE ELEMENT METHOD FOR INTERIOR PROBLEMS

5.5.1. Preliminary Formulation

We proceed to formulate, in general terms, the linear interior field problem solved by the finite element method. Again, the problem is expressed by the operator equation, Equation (5.4.2-1).

$$K\gamma = \mathbf{g}$$

where γ is the exact solution (the field).

The operator K in this equation is a differential operator and is of second degree. Furthermore, the boundary conditions of γ are carried by K. This is done by requiring the domain of K, D_K, to contain only those elements of γ that satisfy those boundary conditions. If the boundary conditions on γ are inhomogenous, D_K is not a linear space and K is not a linear operator. If the boundary conditions on γ are homogenous, then D_K is a linear space and K is a linear operator. (See Section 5.3.)

We develop the general finite element solution to this problem, using the Galerkin approach. As shown in the previous section, we start with Equation (5.4.2-2) to yield

$$\langle K\gamma, \boldsymbol{\alpha}_i \rangle = \langle \mathbf{g}, \boldsymbol{\alpha}_i \rangle, \qquad 1 \leqslant i \leqslant N_a \tag{5.5.1-1}$$

It is advantageous to transform this equation to a new equation by an integration by parts. Examples of this integration by parts are given in Chapters 6 and 10. This new equation has the advantage that its highest degree of differentiation is one. This means, in turn, that the basis functions, $\boldsymbol{\alpha}_i$, need be continuous only over the problem domain. If Equation (5.5.1-1) were approximated directly (with the *second*-degree differentiation in its operator), the $\boldsymbol{\alpha}_i$ would have to be continuous *and* have continuous first derivatives. Putting it another way, the integration by parts allows us to use basis functions with Lagrange interpolation instead of the more complicated Hermite interpolation.

As shown in Chapters 6 and 10, we can represent the integration by parts by the equation

$$\langle K\gamma, \boldsymbol{\alpha}_i \rangle = F(\gamma, \boldsymbol{\alpha}_i) + C(\boldsymbol{\alpha}_i), \qquad 1 \leqslant i \leqslant N_a \tag{5.5.1-2}$$

In this equation, F is a bilinear functional (or bilinear form) (Ref. 5, p. 179), and C is a linear functional that is determined, in part, by certain of the boundary conditions.

One result of the integration by parts is that the problem now has two kinds of boundary conditions—*essential* and *natural* (Ref. 5, p. 333). Suppose that our problem domain is D and that its boundary is ∂D. Then ∂D consists of two parts: ∂E, over which a essential boundary condition is imposed; and ∂N, over which a natural boundary condition is imposed, as shown in Figure 5-1. Basically, the difference between essential and natural boundary conditions is this: the approximate solution, $\boldsymbol{\beta}$, is required at the outset to satisfy the essential boundary conditions (in the same way that γ does) before the system of linear equations is solved. The solution of that linear system then gives a $\boldsymbol{\beta}$ that *approximates* the natural boundary conditions. Examples of essential and natural boundary conditions are given in Chapter 6 and 10. For example, in the problems in Chapter 6, the Dirichlet boundary condition is essential, whereas the Neumann boundary condition is natural. In the problems in

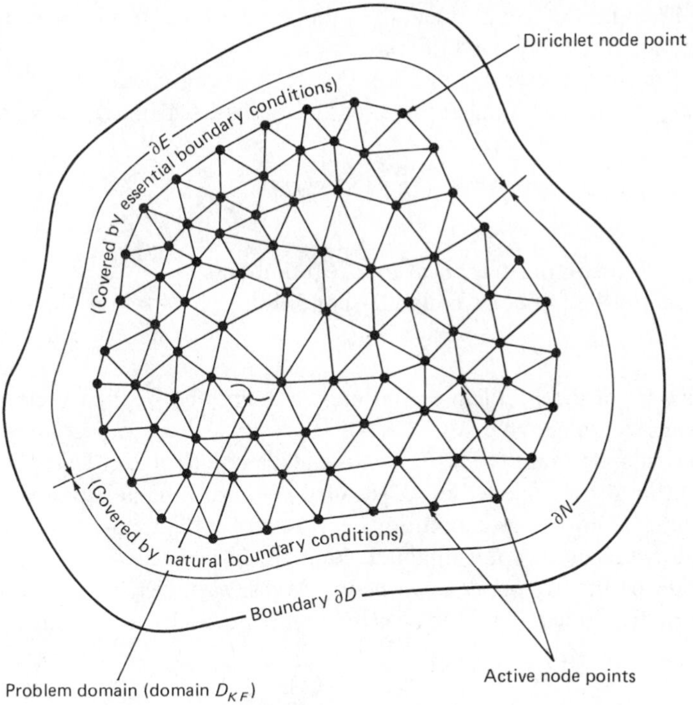

Fig. 5-1. Essential and natural boundary conditions.

this book, the linear functional, C, is determined by the *natural* boundary conditions.

By combining Equations (5.5.1-1) and (5.5.1-2), we have

$$F(\gamma, \boldsymbol{\alpha}_i) = \langle \mathbf{g}, \boldsymbol{\alpha}_i \rangle - C(\boldsymbol{\alpha}_i), \qquad 1 \leqslant i \leqslant N_a \tag{5.5.1-3}$$

We require the approximate solution, $\boldsymbol{\beta}$, to satisfy this same equation:

$$F(\boldsymbol{\beta}, \boldsymbol{\alpha}_i) = \langle \mathbf{g}, \boldsymbol{\alpha}_i \rangle - C(\boldsymbol{\alpha}_i), \qquad 1 \leqslant i \leqslant N_a \tag{5.5.1-4}$$

Section 5.7 shows that this requirement insures that $\boldsymbol{\beta}$ is a linear projection of γ onto a finite dimensional subspace.

5.5.2. Partition of the Approximate Solution, $\boldsymbol{\beta}$

As shown in Figure 5-1 and discussed earlier in this section, a Dirichlet boundary condition (which is essential) is imposed over at least a portion of

∂D, the boundary of the problem domain. We proceed to prescribe the way in which this boundary condition is imposed.

As shown in Chapter 4, Equation (4.2-1), the approximate finite element solution, $\boldsymbol{\beta}$, is given as a linear combination of the basis functions, $\boldsymbol{\alpha}_i$, by

$$\boldsymbol{\beta} = \sum_{i=1}^{N} \beta_i \boldsymbol{\alpha}_i$$

where the β_i are, in most applications, real numbers.

We impose the Dirichlet boundary condition in $\boldsymbol{\beta}$ by taking the following steps:

1. As part of the problem definition, we define the boundary conditions over ∂D. We designate that part of ∂D over which we impose an essential Dirichlet boundary to be ∂E. (∂E could also be *all* of ∂D.) That portion of ∂D that is not included in ∂E is called ∂N and is the portion over which a natural boundary condition is imposed [see Figure 5-1].
2. We call the node points that fall along ∂E, *Dirichlet node points.* All other node points are *active node points.* We say that N_d is the number of Dirichlet node points, and N_a is the number of active node points, so that

$$N = N_a + N_d$$

3. For each Dirichlet node point, i, the β_i equals the Dirichlet boundary condition at that node point. (This works out since, as shown in Chapter 4, the value of $\boldsymbol{\alpha}_i$ at its node point is unity.)

We partition $\boldsymbol{\beta}$ into

$$\boldsymbol{\beta} = \boldsymbol{\beta}_h + \boldsymbol{\beta}_b \qquad (5.5.2\text{-}1)$$

where $\boldsymbol{\beta}_h$ is a linear combination of basis functions that correspond to active node points, and $\boldsymbol{\beta}_b$ is a linear combination of the basis functions of the Dirichlet node points. For convenience, we take the indices i of the active node points to be given by $1 \leqslant i \leqslant N_a$ and the indices j of the Dirichlet node points to be given by $1 + N_a \leqslant j \leqslant N$. We can then partition Equation (4.2-1) into

$$\boldsymbol{\beta}_h = \sum_{i=1}^{N_a} \beta_i \boldsymbol{\alpha}_i \qquad (5.5.2\text{-}2)$$

and

$$\boldsymbol{\beta}_b = \sum_{i=1+N_a}^{N} \beta_i \boldsymbol{\alpha}_i \tag{5.5.2-3}$$

Notice that the coefficients β_i in Equation (5.5.2-3) are just those that were assigned to satisfy the Dirichlet boundary conditions in step 3 above. Then, once the $\boldsymbol{\alpha}_i$ are all known, $\boldsymbol{\beta}_b$ is fixed, at the outset of the problem. After that, it remains only to calculate the β_i that appear in Equation (5.5.2-2).

5.5.3. Formulation for Computation of $\boldsymbol{\beta}_h$

We proceed to use the equations developed earlier in this section to formulate the computation of $\boldsymbol{\beta}_h$. Substituting Equation (5.5.2-1) into Equations (5.5.1-3) and (5.5.1-4) yields

$$F(\boldsymbol{\beta}_h, \boldsymbol{\alpha}_i) = \langle \mathbf{g}, \boldsymbol{\alpha}_i \rangle - F(\boldsymbol{\beta}_b, \boldsymbol{\alpha}_i) - C(\boldsymbol{\alpha}_i) \tag{5.5.3-1}$$

and

$$F((\boldsymbol{\gamma} - \boldsymbol{\beta}_b) - \boldsymbol{\beta}_h, \boldsymbol{\alpha}_i) = 0, \qquad 1 \leqslant i \leqslant N_a \tag{5.5.3-2}$$

When $\boldsymbol{\beta}_h$ in Equation (5.5.3-1) is expanded as a linear combination of basis functions, as in Equation (5.5.2-2), we have the linear system

$$\sum_{j=1}^{N_a} F(\boldsymbol{\alpha}_i, \boldsymbol{\alpha}_j)\boldsymbol{\beta}_j = \langle \mathbf{g}, \boldsymbol{\alpha}_i \rangle - F(\boldsymbol{\beta}_b, \boldsymbol{\alpha}_i) - C(\alpha_i), \qquad 1 \leqslant i \leqslant N_a \tag{5.5.3-3}$$

The heart of the finite element field computation is the solution of this system of linear equations. It is emphasized that all finite element problems discussed in this book (Chapters 6 and 10) conform to this system of equations.

Furthermore, starting with the system of equations in Equation (5.5.3-2), Section 5.7 shows that $\boldsymbol{\beta}_h$ is a *linear projection* of $\boldsymbol{\gamma} - \boldsymbol{\beta}_b$ onto a finite dimensional Hilbert space that has, as a basis, the $\boldsymbol{\alpha}_i$ for $1 \leqslant i \leqslant N_a$.

5.6. INTEGRAL EQUATION METHOD

At least in the general terms used in this chapter, the formulation of the integral equation method is simpler than that of the finite element method. Again, the problem is expressed by the operator equation, Equation (5.4.2-1):

$$K\boldsymbol{\gamma} = \mathbf{g}$$

The operator K in this equation is an integral operator. Typically, K involves an integration of γ over the source problem domain. As exemplified in Chapter 8, any boundary conditions are carried not by K but instead by the right-hand function, $\mathbf{g}$. As a result, D_K is a linear space and K is, in this case, a linear operator. Furthermore, since K is an *integral* operator, an integration by parts is not necessary. Since boundary conditions are not applied to γ, *all* node points in the source problem domain are active node points, so that

$$N_a = N$$

5.6.1. Galerkin Approach to the Integral Equation Method

Applying the Bubnov-Galerkin approach to the integral equation problem, we use Equation (5.4.2-2).

$$\langle K\gamma - \mathbf{g}, \boldsymbol{\alpha}_i \rangle = 0, \qquad 1 \leqslant i \leqslant N_a$$

In this case, the approximate solution, $\boldsymbol{\beta}$, satisfies this same equation:

$$\langle K\boldsymbol{\beta} - \mathbf{g}, \boldsymbol{\alpha}_i \rangle = 0, \qquad 1 \leqslant i \leqslant N_a \tag{5.6.1-1}$$

From these two equations, we have

$$\langle K(\gamma - \boldsymbol{\beta}), \boldsymbol{\alpha}_i \rangle = 0, \qquad 1 \leqslant i \leqslant N_a \tag{5.6.1-2}$$

Again, in Section 5.7, we show that $\boldsymbol{\beta}$ is a projection of γ onto $\mathscr{H}_N$ (that has as a basis $\boldsymbol{\alpha}_i$ for $1 \leqslant i \leqslant N$). When $\boldsymbol{\beta}$ is expanded as a linear combination of the basis functions, $\boldsymbol{\alpha}_i$, as in Equation (4.2-1), Equation (5.6.1-1) becomes

$$\sum_{j=1}^{N} \langle K\boldsymbol{\alpha}_j, \boldsymbol{\alpha}_i \rangle \beta_j = \langle \mathbf{g}, \boldsymbol{\alpha}_i \rangle, \qquad 1 \leqslant i \leqslant N_a \tag{5.6.1-3}$$

The heart of the numerical integral equation solution is the construction and solution of the system of linear equations in Equation (5.6.1-3).

5.6.2. Collocation Approach to the Integral Equation Method

Let P represent the location of any point within the source problem domain. And let P_i represent the location of *node point* i in the source problem domain for $1 \leqslant i \leqslant N_a$. Equation (5.4.2-1) can then be written

$$K\gamma(P) - \mathbf{g}(P) = 0 \tag{5.6.2-1}$$

In particular, we can write

$$K\boldsymbol{\gamma}(P_i) - \mathbf{g}(P_i) = 0, \qquad 1 \leqslant i \leqslant N_a \tag{5.6.2-2}$$

Alternatively, this equation can be written

$$\int_D [K\boldsymbol{\gamma}(P) - \mathbf{g}(P)]\delta(P - P_i)\,dv = 0, \qquad 1 \leqslant i \leqslant N_a \tag{5.6.2-3}$$

where D is the source problem domain and δ is the Dirac delta function. We define the bilinear functional G by

$$G(\mathbf{u}, \mathbf{w}) = \int_D \mathbf{u}\mathbf{w}\,dv \tag{5.6.2-4}$$

Using this bilinear functional, Equation (5.6.2-3) can be rewritten*

$$G(K\boldsymbol{\gamma}(P) - \mathbf{g}(P), \delta(P - P_i)) = 0$$

and if we choose as our basis functions, $\mathbf{p_i}$, the delta functions $\delta(P - P_i)$, this equation becomes

$$G(K\boldsymbol{\gamma}(P) - \mathbf{g}(P), \mathbf{p_i}) = 0, \qquad 1 \leqslant i \leqslant N_a \tag{5.6.2-5}$$

We require the approximate numerical solution, $\boldsymbol{\beta}$, to satisfy Equation (5.6.2-2), and therefore Equation (5.6.2-5), so that

$$K\boldsymbol{\beta}(P_i) - \mathbf{g}(P_i) = 0, \qquad 1 \leqslant i \leqslant N_a \tag{5.6.2-6}$$

and

$$G(K\boldsymbol{\beta}(P) - \mathbf{g}(P), \mathbf{p_i}) = 0, \qquad 1 \leqslant i \leqslant N_a \tag{5.6.2-7}$$

From Equations (5.6.2-5) and (5.6.2-7), we have

$$G(K[\boldsymbol{\gamma}(P) - \boldsymbol{\beta}(P)], \mathbf{p_i}) = 0, \qquad 1 \leqslant i \leqslant N_a \tag{5.6.2-8}$$

and, using this equation, we can show (see Section 5.7) that $\boldsymbol{\beta}(P)$ is a projection of $\boldsymbol{\gamma}(P)$ onto $\mathscr{H}_N$ (the subspace that has, as basis, $\boldsymbol{\alpha}_i$ for $1 \leqslant i \leqslant N_a$).

* For the collocation method, we must use the bilinear functional, G, instead of an inner product. This is because the Dirac delta function, δ, is *not* square integrable, and therefore cannot be an element in a Hilbert space.

If $\boldsymbol{\beta}(P_i)$ is Equation (5.6.2-6) is an expanded linear combination of pulse-type basis functions, by Equation (4.2-1), we have

$$\sum_{j=1}^{N_a} \beta_j K \boldsymbol{\alpha}_j(P_i) = \mathbf{g}(P_i), \qquad 1 \leqslant i \leqslant N_a \tag{5.6.2-9}$$

We solve this linear system for β_j for $1 \leqslant j \leqslant N_a$.

5.7. PROJECTION METHODS

This section shows that any numerical solution of a linear field problem is in fact a linear projection of the exact solution onto a finite dimensional subspace. This is true, whether the approach is Rayleigh-Ritz, Galerkin, or Galerkin-related, and whether the finite element method or the integral equation method is used.

Section 5.7.1 defines mathematically what is meant by a linear projection. Sections 5.7.2 and 5.7.3 then develop a common formulation that can represent either the finite element method or the integral equation method, when applied to linear problems. Finally, Section 5.7.4 uses this formulation and the results of Section 5.7.1 to show that these numerical solutions are linear projections.

5.7.1. Projections

If A and B are linear spaces, then we define $A + B$ to be the set of all elements of the form $\boldsymbol{\alpha} + \boldsymbol{\beta}$, where $\boldsymbol{\alpha}$ is an element of A and $\boldsymbol{\beta}$ is an element of B (Ref. 7, p. 20). Furthermore, we can show that $A + B$ is a linear space. If, in addition, the intersection of A and B is precisely the zero element, that is,

$$A \cap B = \{0\} \tag{5.7.1-1}$$

then we say that A and B are *disjoint*, that $A + B$ is a *direct sum* of A and B, and that $A + B$ is the *direct sum linear space*. Suppose, in addition, that the element γ is in $A + B$, and

$$\gamma = \boldsymbol{\alpha} + \boldsymbol{\beta} \tag{5.7.1-2}$$

with $\boldsymbol{\alpha}$ in A and $\boldsymbol{\beta}$ in B. Then, if Equation (5.7.1-1) holds, we can show that $\boldsymbol{\alpha}$ and $\boldsymbol{\beta}$ are uniquely determined. To do this, let

$$\gamma = \boldsymbol{\alpha}_1 + \boldsymbol{\beta}_1 = \boldsymbol{\alpha}_2 + \boldsymbol{\beta}_2$$

from which we have

$$\boldsymbol{\alpha}_1 - \boldsymbol{\alpha}_2 = \boldsymbol{\beta}_2 - \boldsymbol{\beta}_1 \tag{5.7.1-3}$$

Since $(\boldsymbol{\alpha}_1 - \boldsymbol{\alpha}_2)$ is in A, and $(\boldsymbol{\beta}_1 - \boldsymbol{\beta}_2)$ is in B, we see from Equation (5.7.1-3) that both elements $(\boldsymbol{\alpha}_1 - \boldsymbol{\alpha}_2)$ and $(\boldsymbol{\beta}_1 - \boldsymbol{\beta}_2)$ are in $A \cap B$. From Equation (5.7.1-1), then,

$$\boldsymbol{\alpha}_1 - \boldsymbol{\alpha}_2 = 0$$

$$\boldsymbol{\beta}_1 - \boldsymbol{\beta}_2 = 0 \tag{5.7.1-4}$$

From Equations (5.7.1-4) we see that, for a direct sum, $\boldsymbol{\alpha}$ and $\boldsymbol{\beta}$ are uniquely defined by $\boldsymbol{\gamma}$.

From this uniqueness, we can define a mapping, P, that maps $\boldsymbol{\gamma}$ into $\boldsymbol{\beta}$. We can show that this mapping is a linear transformation,

$$P(\boldsymbol{\gamma}) = \boldsymbol{\beta} \tag{5.7.1-5}$$

We say that P is the *projection operator* of $\boldsymbol{\gamma}$ *along* A *onto* B. The domain of P is the linear space $A + B$, and the range of P is the linear space B. For convenience, we call A the *along subspace* and B the *onto subspace.*

Figure 5-2 shows a very simple example of a projection. In this case, the direct sum linear space, $A + B$, is the plane of the book page, a two-dimensional space. Spaces A and B are one-dimensional spaces, represented by lines A and B on the figure. For example, each element $\boldsymbol{\alpha}$ of the linear space A is a point on line A of the figure. Notice that the intersection of lines A and B (or spaces A and B) is, in fact, the zero element, so that these spaces

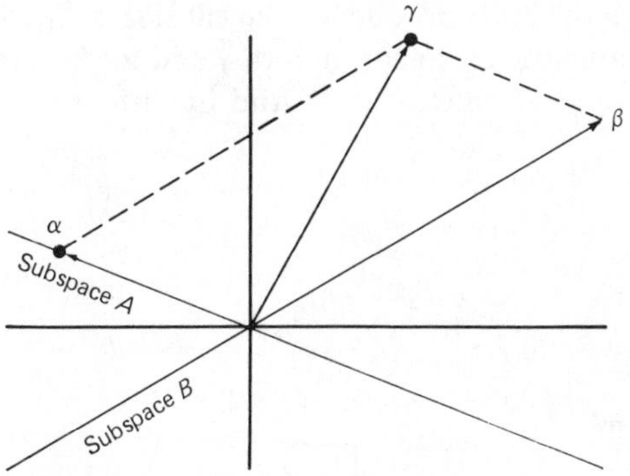

Fig. 5-2. Projection.

are disjoint. The figure shows a point γ in $A + B$ and its projection, $\boldsymbol{\beta}$, into space B.

5.7.2. Expression of Solution of Linear Field Problems in Terms of Bilinear Functionals

The first step in showing that the numerical solutions to linear field problems are linear projections is to show that these solutions can be expressed in terms of bilinear functionals.

From Section 5.6 we see that the solution of all linear field problems by the integral equation method, using all approaches *except* the collocation approach, can be expressed in general by the linear systems

$$\langle K\boldsymbol{\gamma}, \boldsymbol{\alpha}_i \rangle = \langle \mathbf{g}, \boldsymbol{\alpha}_i \rangle, \qquad 1 \leqslant i \leqslant N_a \tag{5.7.2-1}$$

and

$$\langle K\boldsymbol{\beta}, \boldsymbol{\alpha}_i \rangle = \langle \mathbf{g}, \boldsymbol{\alpha}_i \rangle, \qquad 1 \leqslant i \leqslant N_a \tag{5.7.2-2}$$

where

$$\boldsymbol{\beta} = \sum_{j=1}^{N_a} \beta_j \boldsymbol{\alpha}_j \tag{5.7.2-3}$$

and K is a linear operator. Furthermore, these equations hold, regardless of whether the Rayleigh-Ritz or Galerkin or related approaches (except the collocation approach) are used. To adapt these equations to the collocation approach, we need only replace $\boldsymbol{\alpha}_i$ by the Dirac delta function $\delta(P - P_i)$ and the inner product by the bilinear functional, G (as discussed in Section 5.6.2).

Notice that both the inner product on the left side of Equation (5.7.2-1) and the bilinear functional G, are linear in both $\boldsymbol{\gamma}$ and $\boldsymbol{\alpha}_i$. As a result, they can be expressed by the bilinear functional, F. And Equations (5.7.2-1) and (5.6.2-5) become

$$F(\boldsymbol{\gamma}, \boldsymbol{\alpha}_i) = b_i, \qquad 1 \leqslant i \leqslant N_a \tag{5.7.2-4}$$

where*

$$b_i = \langle \mathbf{g}, \boldsymbol{\alpha}_i \rangle, \qquad 1 \leqslant i \leqslant N_a \tag{5.7.2-5}$$

---

* For the collocation method, we have $b_i = G(\mathbf{g}, \mathbf{p}_i)$ from Equation (5.6.2-5).

By similar reasoning, Equation (5.7.2-2) becomes

$$F(\boldsymbol{\beta}, \boldsymbol{\alpha}_i) = b_i, \qquad 1 \leqslant i \leqslant N_a \tag{5.7.2-6}$$

Subtracting Equation (5.7.2-5) from Equation (5.7.2-3), we have

$$F(\boldsymbol{\gamma} - \boldsymbol{\beta}, \boldsymbol{\alpha}_i) = 0 \tag{5.7.2-7}$$

From Section 5.5, we see that if $\boldsymbol{\gamma}$ is replaced by $\boldsymbol{\gamma} - \boldsymbol{\beta}_b$, if $\boldsymbol{\beta}$ is replaced by $\boldsymbol{\beta}_h$, and if we let

$$b_i = \langle \mathbf{g}, \boldsymbol{\alpha}_i \rangle - F(\boldsymbol{\beta}_b, \boldsymbol{\alpha}_i) - C(\boldsymbol{\alpha}_i) \tag{5.7.2-8}$$

then Equations (5.7.2-3) through (5.7.2-7) also apply to the finite element method implemented by the Galerkin or related approaches. (One could also show that these equations apply to the finite element method as implemented by the Rayleigh-Ritz approach.)

Summing up, we see that the solutions to all linear numerical problems, whether they are solved by the finite element method or the integral equation method, as implemented by either the Rayleigh-Ritz or Galerkin or related approaches, can be expressed by Equations (5.7.2-3) through (5.7.2-7).

5.7.3. Numerical Solution of Linear Field Problems

Combining Equations (5.7.2-3) and (5.7.2-6) gives us

$$\sum_{j=1}^{N_a} F(\boldsymbol{\alpha}_j, \boldsymbol{\alpha}_i)\beta_j = b_i, \qquad 1 \leqslant i \leqslant N_a \tag{5.7.3-1}$$

These equations yield the linear system that is solved for a numerical solution to a linear field problem. It is most practical and convenient to put this system of equations into the matrix-vector form

$$M\hat{\boldsymbol{\beta}} = \mathbf{B} \tag{5.7.3-2}$$

In this equation, M is a matrix having as elements

$$m_{ij} = F(\boldsymbol{\alpha}_j, \boldsymbol{\alpha}_i), \qquad 1 \leqslant i, j \leqslant N_a \tag{5.7.3-3}$$

The vectors $\hat{\boldsymbol{\beta}}$ and $\mathbf{B}$ have as elements β_j and b_i, respectively, for $1 \leqslant i, j \leqslant N_a$. The matrix, M, must be nonsingular in order to achieve a unique solution, $\hat{\boldsymbol{\beta}}$. This is usually achieved without difficulty.

Notice that Equation (5.7.3-2) is, in fact, the linear system that is solved in *all* linear field problems that use either the integral equation method or the finite element method.

5.7.4. Solutions to Numerical Field Problems Expressed as Linear Projections

We proceed to establish the circumstances under which the numerical solution to a linear field problem is a linear projection of the exact solution onto a finite-dimensional Hilbert space. We use Equations (5.7.2-3) through (5.7.2-7) for this purpose.

First, we see that the set of all $\boldsymbol{\beta}$ that satisfy Equation (5.7.2-3) is a subspace $\mathscr{H}_{N_a}$ (of dimension N_a) of the Hilbert space $\mathscr{H}$. Next, we define an error, $\mathbf{e}$, as

$$\mathbf{e} = \boldsymbol{\gamma} - \boldsymbol{\beta} \tag{5.7.4-1}$$

so that Equation (5.7.2-7) becomes

$$F(\mathbf{e}, \boldsymbol{\alpha}_i) = 0, \qquad 1 \leqslant i \leqslant N_a \tag{5.7.4-2}$$

Let E be the set of all elements $\mathbf{e}$ that satisfy the N_a of Equation (5.7.4-2). Then E is a linear space and is another subspace of $\mathscr{H}$. In the terminology of Section 5.7.1, E is the *along* subspace, and $\mathscr{H}_{N_a}$ is the *onto* subspace.

From Equation (5.7.4-1), we see that $\boldsymbol{\beta}$ is a projection of $\boldsymbol{\gamma}$ along E and onto $\mathscr{H}_{N_a}$, provided that E and $\mathscr{H}_{N_a}$ are disjoint. Figure 5-3 shows diagrammatically the exact solution, $\boldsymbol{\gamma}$, and the subspace E and $\mathscr{H}_{N_a}$.

We proceed to derive the circumstances in which E and $\mathscr{H}_{N_a}$ are disjoint. To do this, we let $\mathbf{I}$ be an element of $\mathscr{H}$ that is in both E and $\mathscr{H}_{N_a}$. When $\mathbf{I}$

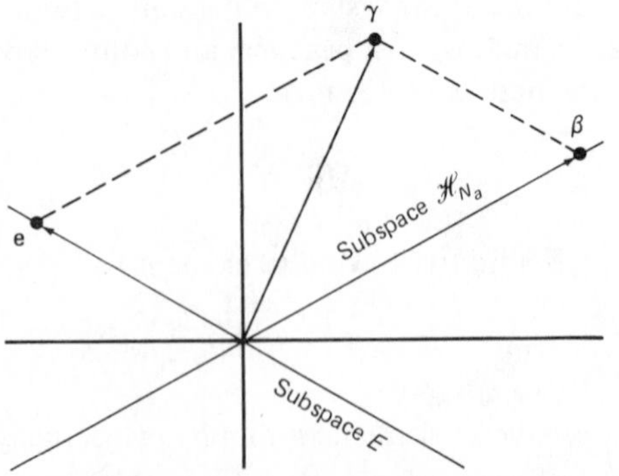

Fig. 5-3. Projection in a field calculation.

must be the zero element, then E and $\mathscr{H}_{N_a}$ are disjoint. Since $\mathbf{I}$ is in E, then, from Equation (5.7.4-2),

$$F(\mathbf{I}, \boldsymbol{\alpha}_i) = 0, \qquad 1 \leqslant i \leqslant N_a \tag{5.7.4-3}$$

Furthermore, since $\mathbf{I}$ is in $\mathscr{H}_{N_a}$, we have from Equation (5.7.2-3) that

$$\mathbf{I} = \sum_{j=1}^{N_a} I_j \boldsymbol{\alpha}_j \tag{5.7.4-4}$$

In this equation, the I_j are real numbers. When these two equations are combined, we have

$$\sum_{j=1}^{N_a} I_j F(\alpha_j, \alpha_i) = 0, \qquad 1 \leqslant i \leqslant N_a \tag{5.7.4-5}$$

Notice that the bilinear functionals in this equation are the same as the elements of the matrix, M, given in Equation (5.7.3-3), so that this equation can be written

$$\sum_{j=1}^{N_a} m_{ij} I_j = 0, \qquad 1 \leqslant i \leqslant N_a \tag{5.7.4-6}$$

From this system of equations, we see that

1. $I_j = 0, \qquad 1 \leqslant j \leqslant N_a$;
2. The element $\mathbf{I}$ of $\mathscr{H}$ is the zero element;
3. The subspaces E and $\mathscr{H}_{N_a}$ are disjoint;
4. $\boldsymbol{\beta}$ is a linear projection of $\boldsymbol{\gamma}$ onto $\mathscr{H}_{N_a}$;

if and only if the matrix M of Equation (5.7.3-2) is nonsingular. This seems quite reasonable, since, by the Kronecker-Kapelli theorem (Ref. 8, p. 55), we know that the matrix-vector Equation (5.7.3-2) has a unique solution if and only if M is nonsingular.

5.8. ORTHOGONAL PROJECTION METHODS

Suppose that the bilinear functional F that was used in the last section happens to be symmetric and positive definite. Then this bilinear functional meets all the requirements of an inner product stated in Equations (5.2.4-1) through (5.2.4-5). That is, this bilinear functional becomes an inner product, so that, if $\mathbf{u}$ and $\mathbf{w}$ are elements of a Hilbert space, $\mathscr{H}$, then

$$F(\mathbf{u}, \mathbf{w}) = \langle \mathbf{u}, \mathbf{w} \rangle_2 \tag{5.8-1}$$

The subscript 2 on the inner product in this equation indicates that this inner product is, in general, not the same as the inner product of the Hilbert space, $\mathscr{H}$, from which the bilinear functional F was originally derived.

Section 5.8.1 shows that, when Equation (5.8-1) holds, then

1. The subspaces E and $\mathscr{H}_{N_a}$ are always disjoint.
2. The approximate solution, $\boldsymbol{\beta}$, is always a linear projection.
3. This projection is an *orthogonal* projection.

Section 5.8.2 shows that an orthogonal projection minimizes the natural norm of the error. It also shows, in a general way, the increase in the natural norm of the error that occurs when a solution is a nonorthogonal projection. It is of interest, therefore, to differentiate between numerical solutions that are orthogonal projections and those that are nonorthogonal projections. Accordingly, Section 5.8.3 differentiates between finite element solutions that are orthogonal projections and those that are nonorthogonal projections. Finally, Section 5.9.4 differentiates between integral equation solutions that are orthogonal projections and those that are nonorthogonal projections.

5.8.1. Solutions to Linear Field Problems Expressed as Orthogonal Projections

We proceed to show that if the bilinear functional F is an inner product, then our linear field problem always has an approximate solution, $\boldsymbol{\beta}$, that is a linear projection. In addition, this linear projection is always an orthogonal projection. The argument given in Section 5.7.4 is repeated, but with an inner product in place of the bilinear functional F. Again, we see that $\boldsymbol{\beta}$ is a projection of $\boldsymbol{\gamma}$ onto $\mathscr{H}_{N_a}$, provided that the subspaces E and $\mathscr{H}_{N_a}$ are disjoint. We prove that if the bilinear functional F is an inner product, then E and $\mathscr{H}_{N_a}$ are always disjoint. From Equation (5.8-1), we see that in this case E is the subspace of elements, $\mathbf{e}$, for which

$$\langle \mathbf{e}, \boldsymbol{\alpha}_i \rangle_2 = 0, \qquad 1 \leqslant i \leqslant N_a \tag{5.8.1-1}$$

We let $\mathbf{I}$ be an element in the intersection of E and $\mathscr{H}_{N_a}$. Since $\mathbf{I}$ is in E, then it obeys Equation (5.8.1-1) and

$$\langle \mathbf{I}, \boldsymbol{\alpha}_i \rangle_2 = 0, \qquad 1 \leqslant i \leqslant N_a \tag{5.8.1-2}$$

But, since **I** is in $\mathscr{H}_{N_a}$, it can be expressed as a linear combination of the basis functions, $\boldsymbol{\alpha}_i$ [as $\boldsymbol{\beta}$ is in Equation (5.7.2-3)], so that

$$\mathbf{I} = \sum_{j=1}^{N_a} I_j \boldsymbol{\alpha}_j \tag{5.8.1-3}$$

where the $\mathbf{I}_j$ are real numbers. Taking the inner product of this equation with **I**, we have

$$\langle \mathbf{I}, \mathbf{I} \rangle_2 = \left\langle \mathbf{I}, \sum_{j=1}^{N_a} I_j \boldsymbol{\alpha}_j \right\rangle_2 = \sum_{j=1}^{N_a} I_j \langle \mathbf{I}, \boldsymbol{\alpha}_j \rangle_2$$

and, with Equation (5.8.1-2),

$$\langle \mathbf{I}, \mathbf{I} \rangle_2 = 0$$

from which

$$\mathbf{I} = 0$$

Since **I** is zero, the subspaces E and $\mathscr{H}_{N_a}$ are disjoint, and $\boldsymbol{\beta}$ is a projection of γ into $\mathscr{H}_{N_a}$.

Referring to Equations (5.7.3-2), (5.7.3-3), and (5.8-1), we see that the matrix M has, in this case, the elements

$$m_{ij} = \langle \boldsymbol{\alpha}_i, \boldsymbol{\alpha}_j \rangle_2 \tag{5.8.1-4}$$

From this equation it is not hard to prove that M is, in this case, symmetric and positive definite. These facts can be used to good advantage in choosing a method for solving the linear system that uses M.

From Equations (5.7.2-7) and (5.8-1), we have

$$\langle \gamma - \boldsymbol{\beta}, \boldsymbol{\alpha}_i \rangle_2 = 0 \tag{5.8.1-5}$$

This equation and Equation (5.7.2-3) show that $\boldsymbol{\beta}$ is, in this case, the *orthogonal projection* of γ onto the finite-dimensional Hilbert space, $\mathscr{H}_{N_a}$ (Ref. 9, p. 16). Figure 5-4 shows, diagrammatically, a computed field as an orthogonal projection. Notice that, in this figure, the approximate solution, $\boldsymbol{\beta}$, and the error, **e**, are *orthogonal* to each other. Comparison of Figures 5-3 and 5-4 shows, intuitively, what is proven in the next subsection, that the orthogonal projection produces the *minimum* error.

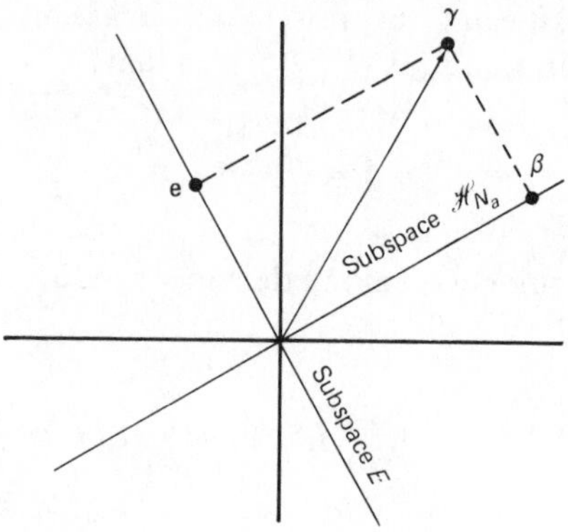

Fig. 5-4. Orthogonal projection in field calculation.

5.8.2. Comparison of Error Norms for Orthogonal and Nonorthogonal Projections

We define a Hilbert space, $\mathscr{H}_2$, as having all the same elements as $\mathscr{H}$ but the inner product $\langle \cdot \rangle_2$ of Equation (5.8-1) as its inner product. Then the natural norm of $\mathscr{H}_2$, $\|\cdot\|_2$, is given by

$$\|\mathbf{u}\|_2 = \langle \mathbf{u}, \mathbf{u} \rangle_2^{1/2} \tag{5.8.2-1}$$

where $\mathbf{u}$ is an element of $\mathscr{H}_2$. Furthermore, we define a finite-dimensional subspace, $\mathscr{H}_{2N_a}$, of $\mathscr{H}_2$ as that subspace that has $\boldsymbol{\alpha}_i$ for $1 \leqslant i \leqslant N_a$ as a basis. Then, the dimension of $\mathscr{H}_{2N_a}$ is N_a. Suppose that $\boldsymbol{\beta}_0$ is an element of $\mathscr{H}_{2N_a}$ that is an orthogonal projection of $\boldsymbol{\gamma}$ onto $\mathscr{H}_{2N_a}$. And suppose that $\boldsymbol{\beta}_n$ is another element of $\mathscr{H}_{2N_a}$ that is a nonorthogonal projection of $\boldsymbol{\gamma}$ onto $\mathscr{H}_{2N_a}$. We proceed to show that

$$\|\boldsymbol{\gamma} - \boldsymbol{\beta}_n\|_2 > \|\boldsymbol{\gamma} - \boldsymbol{\beta}_0\|_2$$

We can certainly express the square of the left side of this inequality by

$$\|\boldsymbol{\gamma} - \boldsymbol{\beta}_n\|_2^2 = \|(\boldsymbol{\gamma} - \boldsymbol{\beta}_0) + (\boldsymbol{\beta}_0 - \boldsymbol{\beta}_n)\|_2^2$$

and, with Equation (5.8.2-1),

$$\|\boldsymbol{\gamma} - \boldsymbol{\beta}_n\|_2^2 = \langle (\boldsymbol{\gamma} - \boldsymbol{\beta}_0) + (\boldsymbol{\beta}_0 - \boldsymbol{\beta}_n), (\boldsymbol{\gamma} - \boldsymbol{\beta}_0) + (\boldsymbol{\beta}_0 - \boldsymbol{\beta}_n) \rangle_2$$

When the inner product of this equation is expanded and we again use Equation (5.8.2-1), we have

$$\|\gamma - \beta_n\|_2^2 = \|\gamma - \beta_0\|_2^2 + 2\langle\gamma - \beta_0, \beta_0 - \beta_n\rangle + \|\beta_0 - \beta_n\|_2^2 \tag{5.8.2-2}$$

Since β_0 is the orthogonal projection of γ onto $\mathscr{H}_{2N_a}$, then the vector $\gamma - \beta_0$ is orthogonal to $\mathscr{H}_{2N_a}$. Since both β_0 and β_n are elements of $\mathscr{H}_{2N_a}$, then the vector $\beta_0 - \beta_n$ lies in $\mathscr{H}_{2N_a}$. Therefore $\beta_0 - \beta_n$ is orthogonal to $\gamma - \beta_0$. The inner product in Equation (5.8.2-2) is zero, so that

$$\|\gamma - \beta_n\|_2^2 = \|\gamma - \beta_0\|_2^2 + \|\beta_0 - \beta_n\|_2^2 \tag{5.8.2-3}$$

From this equation, we see that the error norm for the nonorthogonal projection exceeds the error norm for the orthogonal projection.

5.8.3. Finite Element Solutions That Are Orthogonal Projections

We consider the finite element problems presented in this book. All of these algorithms were developed using integration by parts, in order to reduce the order of integration from two to one. Furthermore, this integration by parts results, in each of these algorithms, in the derivation of a bilinear functional.

Chapter 6 develops the algorithm for the finite element solution to Poisson's equation in the interior problem. As shown there, the bilinear functional that this algorithm develops is symmetric and positive definite, and is therefore an inner product. Thus, the solution to this algorithm is, in fact, an orthogonal projection. The natural norm of this inner product that is minimized by this orthogonal projection is presented in Chapter 6. In Chapter 7, the finite element algorithm presented by Silvester et al. is essentially that for the interior problem presented in Chapter 6. Therefore, the solution to this algorithm is an orthogonal projection. However, the algorithm presented in Chapter 7 for the Wexler-McDonald algorithm contains features that differ from those in the Chapter 6 algorithm. The solution to the Wexler-McDonald algorithm is not an orthogonal projection.

The eddy current problem is solved in Chapter 10 by the finite element method. The bilinear form developed in that algorithm is shown to be symmetric but not positive definite. Therefore, this bilinear form is not an inner product, and the solution is a nonorthogonal projection.

5.8.4. Integral Equation Solutions That Are Orthogonal Projections

Referring to Sections 5.6.1 and 5.6.2, we have

$$\langle K(\gamma - \beta), \mathbf{p}_i\rangle = 0, \qquad 1 \leqslant i \leqslant N_a \tag{5.8.4-1}$$

where, for the Galerkin approach, $\mathbf{p}_i = \boldsymbol{\alpha}_i$, and, for the collocation approach, $\mathbf{p}_i$ is a delta function.

When the collocation approach is used, we clearly have a solution that is a nonorthogonal projection.

When the Galerkin approach is used, however, we can have an orthogonal projection, depending upon the linear operator K. If K is a self-adjoint and positive definite, then we see that we can define a new inner product with a subscript 2, given by

$$\langle K\mathbf{u}, \mathbf{w} \rangle = \langle \mathbf{u}, \mathbf{w} \rangle_2$$

where $\mathbf{u}$ and $\mathbf{w}$ are elements of the Hilbert space $\mathscr{H}$. This equation and Equation (5.8.4-1) (letting $\mathbf{p}_i = \boldsymbol{\alpha}_i$) give us

$$\langle \boldsymbol{\gamma} - \boldsymbol{\beta}, \boldsymbol{\alpha}_i \rangle_2 = 0, \qquad 1 \leqslant i \leqslant N_a \tag{5.8.4-2}$$

We define a new Hilbert space $\mathscr{H}_2$ that has, as its inner product, the inner product $\langle \cdot \rangle_2$, of this equation. Since $\boldsymbol{\beta}$ is a linear combination of the basis functions $\boldsymbol{\alpha}_i$ [as in Equation (5.7.2-3)], this equation satisfies the requirements for an orthogonal projection in $\mathscr{H}_2$.

Summing up, if we solve an integral equation field problem using the Galerkin approach, and if our integral operator is self-adjoint and positive definite, then we can find a Hilbert space $\mathscr{H}_2$, in which the solution is an orthogonal projection. For this reason, the integral operators discussed in Chapter 8 are identified as to whether or not they are self-adjoint and positive definite.

REFERENCES

1. Stakgold, Ivan. *Boundary Value Problems of Mathematical Physics*, Vol. 1. New York: The Macmillan Company, 1967.
2. McDonald, B., Friedman, M., and Wexler, A., "Variational Solution of Integral Equations," *IEEE Transactions on Microwave Theory and Techniques*, March 1974, Vol. MTT-22, No. 3.
3. Wexler, Alvin. *Finite Elements for Technologists.* Winnipeg: Electrical Engineering Department, University of Manitoba, 1974.
4. Mikhlin, S. G. *Variational Methods in Mathematical Physics.* New York: The Macmillan Company, 1964.
5. Milne, R. D., "Applied Functional Analysis." Boston: Pitman Advanced Publishing Program, 1980.
6. Harrington, Roger, F. *Field Computation by Moment Methods.* New York: The Macmillan Company, 1968.
7. Nering, Evan D. *Linear Algebra and Matrix Theory.* New York: John Wiley & Sons, 1963.
8. Shilov, Georgi. *Linear Spaces.* Englewood Cliffs, N.J.: Prentice-Hall, Inc., 1961.
9. Mitchell, A. R., and Wait, R. *The Finite Element Method in Partial Differential Equations.* New York: John Wiley & Sons, 1977.

6
FINITE ELEMENT METHOD FOR INTERIOR PROBLEMS

6.1. INTRODUCTION

The finite element method has been used for several decades to obtain numerical solutions to different types of problems that arise in science and engineering. These problems include stress and strain in solid bodies, fluid flow, and heat flow. As a result, a number of books have been written about the application of the finite element method to these problems. In addition, several books have been written about the finite element method itself (Refs. 1, 2, 3, 4). And at least one book has as its subject the application of the finite element method to the computation of electric and magnetic fields (Ref. 5).

This chapter presents the finite element method as it is applied to static potential, linear interior problems. As we see in this chapter and the next, the finite element method is much more easily applied to interior problems than it is to exterior problems (Chapter 7 deals with exterior problems.)

A field computation by the finite element method (for either interior or exterior problems) is composed of the following steps:

1. Define the problem (including the problem domain) as discussed in Chapter 3.
2. Cover the problem with node points and finite elements; over these, define shape functions, basis functions, and the finite element subspace, S_N, as discussed in Chapter 4.
3. Construct a system of equations in terms of the field to be computed.
4. Solve this system of equations for the field.

Section 6.2 develops, in general, the system of equations in step 3 above, in terms of the problem definition of step 1 and the basis functions of step 2. It also derives a certain error norm and shows that the finite element method minimizes this error norm. Section 6.3 develops formulas that aid in the construction of the system of equations, for certain specific finite elements and shape functions. Finally, Section 6.4 presents and solves a simple sample problem to aid the reader in seeing how the various steps of the finite element method fit together.

6.2. FORMULATION OF FINITE ELEMENT METHOD FOR INTERIOR PROBLEMS

6.2.1. Basic Formulation of Interior Problem

The finite element formulation for interior field problems given below applies to deterministic static and certain quasi-static problems. It does not apply to eigenproblems, which are beyond the scope of this book. We assume that the exact field is represented by the scalar potential, γ, within some region, D, that has a boundary, ∂D, as shown in Figure 6-1. Throughout D, γ satisfies the equation

$$-\nabla\cdot(\nu\nabla\gamma) - f = 0 \tag{6.2.1-1}$$

The boundary ∂D consists in part of ∂E, over which the Dirichlet boundary condition

$$\gamma = g \tag{6.2.1-2}$$

is prescribed, where g is a function of position over ∂E. Over the remainder of the boundary, designated ∂N, the mixed boundary condition applies:

$$(\nu\nabla\gamma)\cdot\mathbf{n} + \sigma\gamma = h \tag{6.2.1-3}$$

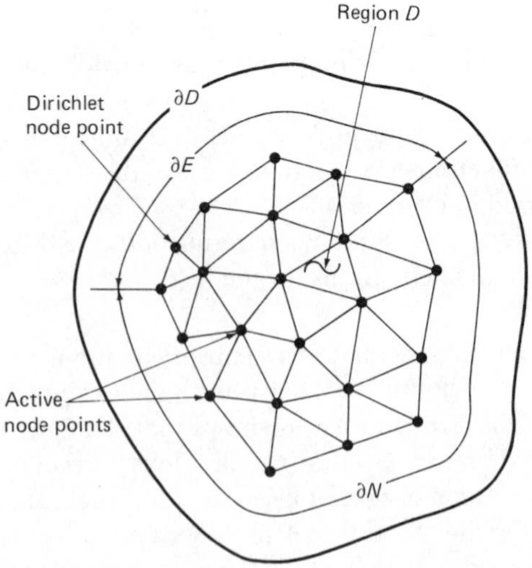

Fig. 6-1. Configuration for finite element solution to field problem.

In this equation, v is either a symmetric positive-definite tensor (such as the permeability tensor or the permittivity tensor) or a positive scalar (such as the scalar permittivity or permeability for an isotropic medium). The vector $\mathbf{n}$ is a unit normal directed outward from the region, and σ, as shown in Chapter 3, is a non-negative scalar. Again, h is a function of position over ∂N. For the formulation given below, we make the Dirichlet boundary condition an essential boundary condition and show later that the mixed boundary condition is, in fact, a natural boundary condition. Then, we have an essential (and Dirichlet) boundary condition over ∂E and a natural (and mixed) boundary condition over ∂N.

As in Section 5.5.2, the finite difference approximate solution, $\boldsymbol{\beta}_F$, is divided into

$$\boldsymbol{\beta}_F = \boldsymbol{\beta}_h + \boldsymbol{\beta}_b \tag{6.2.1-4}$$

where $\boldsymbol{\beta}_F$ and $\boldsymbol{\beta}_b$ satisfy the Dirichlet (and essential) boundary condition at the Dirichlet node points, and $\boldsymbol{\beta}_h$ equals zero over ∂E. Again, neither $\boldsymbol{\beta}_F$, $\boldsymbol{\beta}_b$, nor $\boldsymbol{\beta}_h$ is required, at the outset, to satisfy the mixed boundary condition over ∂N. The computation itself will take care of that. Again, since

$$\boldsymbol{\beta}_F = \sum_{i=1}^{N} \beta_i \boldsymbol{\alpha}_i \tag{6.2.1-5}$$

then

$$\boldsymbol{\beta}_h = \sum_{i=1}^{N_a} \beta_i \boldsymbol{\alpha}_i \tag{6.2.1-6}$$

and

$$\boldsymbol{\beta}_b = \sum_{i=1+N_a}^{N} \beta_i \boldsymbol{\alpha}_i \tag{6.2.1-7}$$

In Equation (6.2.1-6), the basis functions of the active node points, $\boldsymbol{\alpha}_i$, $1 \leqslant i \leqslant N_a$ (as well as $\boldsymbol{\beta}_h$), all equal zero over ∂E. In Equation (6.2.1-7), the $\boldsymbol{\alpha}_i$ for $1 + N_a \leqslant i \leqslant N$ are linearly independent over ∂E, and their coefficients, β_i, are fixed at values such that $\boldsymbol{\beta}_b$ meets the Dirichlet boundary condition at the Dirichlet node points. In this way, $\boldsymbol{\beta}_b$ is fixed by the Dirichlet boundary condition at the outset of the computation. As in Section 5.5, we have only to compute β_i for $1 \leqslant i \leqslant N_a$.

We use Equation (6.2.1-1) to write

$$\int_D [-\nabla \cdot (v \nabla \gamma) - f] \boldsymbol{\alpha}_i \, dv, \qquad 1 \leqslant i \leqslant N_a \tag{6.2.1-8}$$

Notice that the integrand in Equation (6.2.1-8) contains a second-order differential operator. Now, we can use a simpler set of basis functions, $\boldsymbol{\alpha}_i$, for $1 \leqslant i \leqslant N$, if, instead, we have a first-order differential operator. To do this, we use the divergence theorem, as follows, to perform the integration by parts discussed in Section 5.5.1.

$$\int_D \boldsymbol{\alpha}_i \nabla \cdot (\nu \nabla \gamma)\, dv = \int_{\partial D} \mathbf{n} \cdot (\boldsymbol{\alpha}_i \nu \nabla \gamma)\, dS - \int_D \nabla \boldsymbol{\alpha}_i \cdot (\nu \nabla \gamma)\, dv, \qquad 1 \leqslant i \leqslant N_a \tag{6.2.1-9}$$

Since $\boldsymbol{\alpha}_i$ for $1 \leqslant i \leqslant N_a$ equals zero over ∂E, then the range of integration of the first integral on the right side of Equation (6.2.1-9) can be reduced from ∂D to ∂N. Doing this and substituting Equation (6.2.1-3) into this integral gives us

$$\int_D \boldsymbol{\alpha}_i \nabla \cdot (\nu \nabla \gamma)\, dv = \int_{\partial N} \boldsymbol{\alpha}_i (h - \sigma\gamma)\, dS - \int_D \nabla \boldsymbol{\alpha}_i \cdot (\nu \nabla \gamma)\, dv, \qquad 1 \leqslant i \leqslant N_a \tag{6.2.1-10}$$

One can see that this equation is a particular example of Equation (5.5.1-2), where

$$\langle K\gamma, \boldsymbol{\alpha}_i \rangle = \langle -\nabla \cdot (\nu \nabla \gamma), \boldsymbol{\alpha}_i \rangle = \int_D -\boldsymbol{\alpha}_i \nabla \cdot (\nu \nabla \gamma)\, dv \tag{6.2.1-11}$$

$$F(\gamma, \boldsymbol{\alpha}_i) = \int_D (\nu \nabla \gamma) \cdot \nabla \boldsymbol{\alpha}_i\, dv + \int_{\partial N} \sigma \gamma \boldsymbol{\alpha}_i\, dS \tag{6.2.1-12}$$

and

$$C(\boldsymbol{\alpha}_i) = -\int_{\partial N} h \boldsymbol{\alpha}_i\, dS \tag{6.2.1-13}$$

Combining Equations (6.2.1-8) and (6.2.1-10) yields the final equation in γ,

$$\int_D [\nabla\boldsymbol{\alpha}_i \cdot (\nu\nabla\gamma) - \boldsymbol{\alpha}_i f]\, dv + \int_{\partial N} \boldsymbol{\alpha}_i(\sigma\gamma - h)\, dS = 0, \qquad 1 \leqslant i \leqslant N_a \tag{6.2.1-14}$$

If Equation (6.2.1-14) is used, along with the definition of the bilinear functional, F, given in Equation (6.2.1-12), we have

$$F(\gamma, \boldsymbol{\alpha}_i) = \int_D f\boldsymbol{\alpha}_i\, dv + \int_{\partial N} h\boldsymbol{\alpha}_i\, dS, \qquad 1 \leqslant i \leqslant N_a \tag{6.2.1-15}$$

This, it turns out, is the basic equation from which our numerical solution is developed.

6.2.2. Proof that F is an Inner Product

Before proceeding, it is convenient to prove that the bilinear functional, F, is an inner product. This fact is necessary later when we demonstrate that the algorithm developed in this section is an orthogonal projection.

To begin, we construct a Hilbert space, $\mathscr{H}_2$, that fits our finite element problem. We then show that, if the arguments $\mathbf{u}$ and $\mathbf{w}$ of $F(\mathbf{u}, \mathbf{w})$ are elements of $\mathscr{H}_2$, then F is an inner product. We then choose this as the inner product of $\mathscr{H}_2$.

As elements of $\mathscr{H}_2$, we choose all continuous functions, defined over the problem domain, D, that are zero at the Dirichlet node points on ∂E. (Here, ∂E is that portion of ∂D over which a Dirichlet boundary condition is imposed, as discussed in Section 5.5) As stated in Chapter 5, we assume that ∂E is nonzero in any interior finite element problem. Since γ and $\boldsymbol{\beta}_F$ satisfy the same Dirichlet boundary conditions, at the Dirichlet node points, the function $(\gamma - \boldsymbol{\beta}_F)$ is an element of $\mathscr{H}_2$. In the same way, $(\gamma - \boldsymbol{\beta}_b)$ is an element of $\mathscr{H}_2$. The basis functions of the active node points $\boldsymbol{\alpha}_i$, for $1 \leqslant i \leqslant N_a$, are all zero at the Dirichlet node points. Then, these basis functions are all elements of $\mathscr{H}_2$. Finally, since $\boldsymbol{\beta}_h$ is a linear combination of these basis functions, it is also an element of $\mathscr{H}_2$.

It becomes convenient to define a finite-dimensional subspace of $\mathscr{H}_2$, $\mathscr{H}_{N_a}$, which has as a basis, the N_a, basis functions of the active node points. Then $\boldsymbol{\beta}_h$ is an element of $\mathscr{H}_{N_a}$.

As we see in Section 5.2, a bilinear functional that is symmetric and positive definite is an inner product. We start with the definition of F given in Equation (6.2.1-12) and show that F has these properties. First we recall that ν is either a symmetric positive definite tensor or a positive scalar. It is easy to see, then, that F is symmetric, that is, that

$$F(\mathbf{u}, \mathbf{w}) = F(\mathbf{w}, \mathbf{u}) \tag{6.2.2-1}$$

It is more difficult to show that F is positive definite. To do this, we must show that, for any nonzero element $\mathbf{u}$ in $\mathscr{H}_2$, the expression

$$\int_D (\nu \nabla \mathbf{u}) \cdot \nabla \mathbf{u} \, dv + \int_{\partial N} \sigma \mathbf{u}^2 \, dS$$

is positive. Suppose, then, that $\mathbf{u}$ is nonzero. Then, since $\mathbf{u}$ must be zero at the Dirichlet node points and nonzero at some other point in D, then $\nabla \mathbf{u}$ must be nonzero in D. From this, and the fact that ν is either a positive scalar or a positive definite tensor, we see that the first integral in the above expression must be positive. We assume that σ (which is part of a mixed boundary condition) must be non-negative. Therefore, for a nonzero $\mathbf{u}$ in $\mathscr{H}_2$, the above expression must be positive and F must be positive definite.

Since F is both symmetric and positive definite, it is an inner product. Specifically it is the inner product of $\mathscr{H}_2$, designated by the subscript 2, so that, for any elements $\mathbf{u}$ and $\mathbf{w}$ of $\mathscr{H}_2$,

$$F(\mathbf{u}, \mathbf{w}) = \langle \mathbf{u}, \mathbf{w} \rangle_2 \tag{6.2.2-2}$$

When this equation is substituted into Equation (6.2.1-12),

$$\langle \mathbf{u}, \mathbf{w} \rangle_2 = \int_D \nu \nabla \mathbf{u} \cdot \nabla \mathbf{w} \, dv + \int_{\partial N} \sigma \mathbf{u} \mathbf{w} \, dS \tag{6.2.2-3}$$

which defines this inner product.

6.2.3. Algorithm for Computing $\boldsymbol{\beta}_h$

As in Section 5.5, we make the finite element approximate field, $\boldsymbol{\beta}_F$, satisfy Equation (6.2.1-15) (as γ does) so that

$$F(\boldsymbol{\beta}_F, \boldsymbol{\alpha}_i) = \int_D f \boldsymbol{\alpha}_i \, dv + \int_{\partial N} h \boldsymbol{\alpha}_i \, dS, \qquad 1 \leqslant i \leqslant N_a \tag{6.2.3-1}$$

Using Equation (6.2.1-4) (that expresses $\boldsymbol{\beta}_F$ as the sum of $\boldsymbol{\beta}_b$ and $\boldsymbol{\beta}_h$), and noting that $\boldsymbol{\beta}_b$ is fixed at the outset of the computation, we move it to the right side. Then, noting that $\boldsymbol{\beta}_h$ is in $\mathscr{H}_2$, we have

$$F(\boldsymbol{\beta}_h, \boldsymbol{\alpha}_i) = \langle \boldsymbol{\beta}_h, \boldsymbol{\alpha}_i \rangle_2 = b_i, \qquad 1 \leqslant i \leqslant N_a \tag{6.2.3-2}$$

where

$$b_i = \int_D f\alpha_i \, dv + \int_{\partial N} h\alpha_i \, dS - F(\boldsymbol{\beta}_b, \alpha_i), \qquad 1 \leqslant i \leqslant N_a$$

Furthermore, expressing $\boldsymbol{\beta}_b$ by Equation (6.2.1-7) (as a linear combination of the basis functions of the Dirichlet node points), we have

$$b_i = \int_D f\alpha_i \, dv + \int_{\partial N} h\alpha_i \, dS - \sum_{j=1+N_a}^{N} \beta_j F(\alpha_j, \alpha_i), \qquad 1 \leqslant i \leqslant N_a \qquad (6.2.3\text{-}3)$$

Then, b_i can be calculated as the outset of the problem (with the aid of equations developed in Section 6.3).

When Equation (6.2.1-6) is substituted into Equation (6.2.3-2), we see that the system of equations that we solve is given by

$$\sum_{j=1}^{N_a} \langle \alpha_j, \alpha_i \rangle_2 \beta_j = b_i, \qquad 1 \leqslant i \leqslant N_a \qquad (6.2.3\text{-}4)$$

And the matrix, M, used in this linear system solution has elements, m_{ij}, given by

$$m_{ij} = \langle \alpha_j, \alpha_i \rangle_2, \qquad 1 \leqslant i, j \leqslant N_a \qquad (6.2.3\text{-}5)$$

Using this equation and the definition of this inner product given by Equation (6.2.2-3) gives us

$$m_{ij} = \langle \alpha_j, \alpha_i \rangle_2 = F(\alpha_j, \alpha_i) = \int_D (v\nabla\alpha_j)\cdot\nabla\alpha_i \, dv + \int_{\partial N} \sigma\alpha_j\alpha_i \, dS, \quad 1 \leqslant i, j \leqslant N_a \qquad (6.2.3\text{-}6)$$

This is the equation that is used to compute these matrix elements.

6.2.4. Demonstration that $\boldsymbol{\beta}_h$ is an Orthogonal Projection

Now $\boldsymbol{\beta}_h$ is shown to be the orthogonal projection of $(\gamma - \boldsymbol{\beta}_b)$ onto $\mathscr{H}_{N_a}$. If we subtract Equation (6.2.3-1) from Equation (6.2.1-15),

$$F(\gamma - \boldsymbol{\beta}_F, \alpha_i) = \langle \gamma - \boldsymbol{\beta}_F, \alpha_i \rangle_2 = \langle (\gamma - \boldsymbol{\beta}_b) - \boldsymbol{\beta}_h, \alpha_i \rangle_2 = 0, \qquad 1 \leqslant i \leqslant N_a \qquad (6.2.4\text{-}1)$$

From this set of N_a equations we see that $(\gamma - \boldsymbol{\beta}_b) - \boldsymbol{\beta}_h$ is orthogonal to $\mathscr{H}_{N_a}$. Since $\boldsymbol{\beta}_h$ is an element of $\mathscr{H}_{N_a}$, we see that $\boldsymbol{\beta}_h$ is the orthogonal projection of

$\boldsymbol{\gamma} - \boldsymbol{\beta}_b$ onto $\mathscr{H}_{N_a}$. (Orthogonal projections are discussed in Section 5.8 of Chapter 5.)

Section 5.8 also shows that this orthogonal projection, in a sense, minimizes the error, $\mathbf{e}$, given by

$$\mathbf{e} = \boldsymbol{\gamma} - \boldsymbol{\beta} = (\boldsymbol{\gamma} - \boldsymbol{\beta}_b) - \boldsymbol{\beta}_h$$

Specifically, the orthogonal projection insures that $\boldsymbol{\beta}_h$ is that element of $\mathscr{H}_{N_a}$ for which the natural norm of $\mathbf{e}$ is minimized. Using Equation (6.2.2-3), we see that the norm of $\mathbf{e}$ that is minimized is given by

$$\|\mathbf{e}\| = \sqrt{\langle \mathbf{e}, \mathbf{e} \rangle_2} = \sqrt{\int_D (\nu \nabla \mathbf{e}) \cdot \nabla \mathbf{e} \, dv + \int_{\partial N} \sigma \mathbf{e}^2 \, dS} \qquad (6.2.4\text{-}2)$$

From this equation, one can see the mechanism by which this algorithm minimizes the error. For simplicity, suppose that a mixed boundary condition is not used, since in this case σ and the second integral in the above equation equal zero. In this situation, the gradient of $\mathbf{e}$ is minimized over the problem domain, D, according to this equation. We can think of the error $\mathbf{e}$ in the following way:

1. At each Dirichlet node point, $\mathbf{e}$ is zero.
2. At any point in D the error equals the line integral of ∇e to that point, from a Dirichlet node point.

6.3. COMPUTATION OF LINEAR SYSTEM FOR FINITE ELEMENT METHOD

The objective of this section is to help the reader to compute the terms b_i that appear in Equation (6.2.3-3) and the inner products and bilinear functionals $\langle \boldsymbol{\alpha}_j, \boldsymbol{\alpha}_i \rangle$ and $F(\boldsymbol{\alpha}_j, \boldsymbol{\alpha}_i)$ that appear in this equation and in Equation (6.2.3-4). Section 6.3.1 discusses this inner product, and the terms that appear in Equation (6.2.3-3), to provide the reader with a general "feel" for these terms. The remaining subsections of this section present analytically derived formulations that aid in computing this inner product for certain commonly used finite elements and shape functions.

6.3.1. Preliminary Considerations

Here we consider the integrals that appear in Equations (6.2.3-3) (for b_i) and (6.2.3-6) (for the inner product).

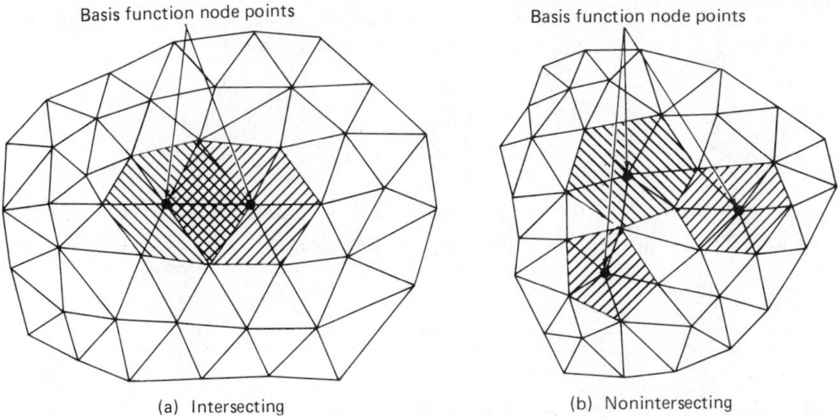

Fig. 6-2. Intersecting and non-intersecting basis functions.

First, since these equations contain the gradients of the basis functions, we must use pyramid-type basis functions [see Figure 4-1b]. [A pulse-type basis function, as in Figure 4-1a, has a gradient that is zero inside its finite element and infinite on the finite element boundaries.]

Second, most of the terms of these equations are zero most of the time. This results largely from the fact that all of the basis functions, $\boldsymbol{\alpha}_i$, used in this book have *local support*. (This property of local support was discussed in Chapter 4.) Those finite elements over which a basis function is nonzero are called its support. For the basis functions in this book, this support includes only those finite elements that contain the basis function's node point. [See Figure 4-1.] If the support of basis function $\boldsymbol{\alpha}_i$ and the support of basis function $\boldsymbol{\alpha}_j$ share one or more finite elements in common [as shown in Figure 6-2a], then we say that these basis functions are *intersecting*. If the supports of two such basis functions do not share any finite elements in common, then these basis functions are *nonintersecting* [as shown in Figure 6-2b]. As we can see, only basis functions that happen to be neighbors of each other are intersecting. We see that the integral

$$\int_D \nabla\boldsymbol{\alpha}_i \cdot (\nu \nabla \boldsymbol{\alpha}_j)\, dv$$

that appears in Equation (6.2.3-6) is nonzero if and only if $\boldsymbol{\alpha}_i$ and $\boldsymbol{\alpha}_j$ are intersecting. The integral

$$\int_D \boldsymbol{\alpha}_i f\, dv$$

can be nonzero only if f is nonzero over the support of $\boldsymbol{\alpha}_i$, but for most of our problems, f is zero over the entire problem domain, D. The integral

$$\int_{\partial N} \boldsymbol{\alpha}_i h \, dS$$

can be nonzero only if the support of $\boldsymbol{\alpha}_i$ and ∂N intersect, and then only if h is nonzero over that intersection. Finally, the integral

$$\int_{\partial N} \sigma \boldsymbol{\alpha}_i \boldsymbol{\alpha}_j \, dS$$

can be nonzero only if the supports of $\boldsymbol{\alpha}_i$, $\boldsymbol{\alpha}_j$, and ∂N intersect and σ is nonzero over that intersection. In most of our problems, we use only Dirichlet and Neumann boundary conditions, and σ is zero.

We can see from the above that if N_a is large—say, of the order of 100 or more (which happens in most practical problems)—then for most pairs of integers i, j, for $1 \leqslant i, j \leqslant N_a$, the quantity m_{ij} is equal to zero. Putting it another way, the vast majority of the N_a^2 elements of the matrix M equal zero. We say that such a matrix is *sparse*. In general, the matrices generated when the finite element method is applied to field problems are sparse.

Finally, we decide which of the integrals in Equations (6.2.3-3) and (6.2.3-6) should be worked out, analytically, for certain specific basis functions, once and for all. The three integrals that involve the functions f, h, and σ depend upon the specific spatial variations that these functions take over R and over ∂N. Furthermore, these integrals are zero, in most problems.

On the other hand, we encounter a large number of problems in which the medium is isotropic, and v is a scalar that is constant over each finite element. We develop a general formulation for this situation. Suppose now that $n(i,j)$ is the total number of finite elements in the intersection of the supports of $\boldsymbol{\alpha}_i$ and $\boldsymbol{\alpha}_j$. For example, in Figure 6-2a, $n(i,j) = 2$. Let k be a local index for each finite element in this intersection. Then we can see that, since v is constant over each finite element,

$$\int_D \nabla \boldsymbol{\alpha}_i \cdot (v \boldsymbol{\alpha}_j) \, dv = \sum_{k=1}^{n(i,j)} v_k \int_{A_k} \nabla \boldsymbol{\alpha}_i \cdot \nabla \boldsymbol{\alpha}_j \, dv \qquad (6.3.1\text{-}1)$$

where A_k is the area (or volume) of finite element k and v_k is the value of v over A_k.

We can see that $\boldsymbol{\alpha}_i$ and $\boldsymbol{\alpha}_j$, as used in Equation (6.3.1-1), really constitute only the shape functions of which $\boldsymbol{\alpha}_i$ and $\boldsymbol{\alpha}_j$ are composed over the finite element A_k. If p_i and p_j are these shape functions, defined over the domain A_k, then for any point p in A_k,

$$p_i(p) = \boldsymbol{\alpha}_i(p)$$

$$p_j(p) = \boldsymbol{\alpha}_j(p)$$

so that

$$\int_{A_k} \nabla\boldsymbol{\alpha}_i \cdot \nabla\boldsymbol{\alpha}_j \, dv = \int_{A_k} \nabla p_i \cdot \nabla p_j \, dv \tag{6.3.1-2}$$

It is convenient to define a bilinear functional, Γ, by

$$\Gamma(\mathbf{u}, \mathbf{w}) = \int_{A_\Gamma} \nabla\mathbf{u} \cdot \nabla\mathbf{w} \, dv \tag{6.3.1-3}$$

where functions $\mathbf{u}$ and $\mathbf{w}$ are both shape functions defined over the same domain, A_Γ. For those problems in which ν is isotropic and uniform over each element, we have, from Equations (6.3.1-1), (6.3.1-2), and (6.3.1-3),

$$\int \nabla\boldsymbol{\alpha}_i \cdot (\nu \nabla\boldsymbol{\alpha}_j) \, dv = \sum_{k=1}^{n(i,j)} \nu_k \Gamma(p_i, p_j)$$

and with Equation (6.2.3-6),

$$m_{ij} = \sum_{k=1}^{n(i,j)} \nu_k \Gamma(p_i, p_j) + \int_{\partial N} \sigma \boldsymbol{\alpha}_i \boldsymbol{\alpha}_j \, dS \tag{6.3.1-4}$$

In succeeding subsections, the formulas for Γ are worked out for a number of shape functions of different dimensions, orders, and shapes that are taken from Chapter 4.

6.3.2. Computation of Γ for Linear Shape Function over a Triangular Finite Element

We proceed to work out Γ for the linear shape functions over a triangle given in Section (4.4.2). In general, for any two-dimensional shape functions that are functions of the Cartesian coordinates x and y, we have, from Equation (6.3.1-3), that

$$\Gamma(\mathbf{u}, \mathbf{w}) = \int_{A_\Gamma} \nabla\mathbf{u}(x, y) \cdot \nabla\mathbf{w}(x, y)\, da$$

$$= \int_{A_\Gamma} \left[\frac{\partial \mathbf{u}(x, y)}{\partial x} \frac{\partial \mathbf{w}(x, y)}{\partial x} + \frac{\partial \mathbf{u}(x, y)}{\partial y} \frac{\partial \mathbf{w}(x, y)}{\partial y} \right] da \qquad (6.3.2\text{-}1)$$

From Equations (4.4.2-5), (4.4.2-6), (4.4.2-7), and (6.3.2-1), we have

$$\Gamma(p_1^1, p_1^1) = \frac{1}{4A_\Gamma} [(y_2 - y_3)^2 + (x_2 - x_3)^2] \qquad (6.3.2\text{-}2)$$

$$\Gamma(p_2^1, p_2^1) = \frac{1}{4A_\Gamma} [(y_3 - y_1)^2 + (x_3 - x_1)^2] \qquad (6.3.2\text{-}3)$$

$$\Gamma(p_3^1, p_3^1) = \frac{1}{4A_\Gamma} [(y_1 - y_2)^2 + (x_1 - x_2)^2] \qquad (6.3.2\text{-}4)$$

$$\Gamma(p_1^1, p_2^1) = \frac{1}{4A_\Gamma} [(y_2 - y_3)(y_3 - y_1) + (x_2 - x_3)(x_3 - x_1)] \qquad (6.3.2\text{-}5)$$

$$\Gamma(p_1^1, p_3^1) = \frac{1}{4A_\Gamma} [(y_2 - y_3)(y_1 - y_2) + (x_2 - x_3)(x_1 - x_2)] \qquad (6.3.2\text{-}6)$$

$$\Gamma(p_2^1, p_3^1) = \frac{1}{4A_\Gamma} [(y_3 - y_1)(y_1 - y_2) + (x_3 - x_1)(x_1 - x_2)] \qquad (6.3.2\text{-}7)$$

6.3.3. Computation of Γ for a Higher Order Shape Function over a Triangular Finite Element

In Subsection 4.4.5, the higher order shape function over a triangle was constructed as a function of the linear shape function. Furthermore, as shown in Subsection 4.4.3, these linear shape functions are the same as the triangular area coordinates. Keeping this in mind, Equation (4.4.5-6) becomes

$$\tau_{ijk}^{(M)} = \ell_i^i(M\Lambda_1)\ell_j^j(M\Lambda_2)\ell_k^k(M\Lambda_3) \qquad (6.3.3\text{-}1)$$

The triangular area coordinates Λ_1, Λ_2, and Λ_3 are used in this equation in place of the linear shape functions over a triangle, p_i^1, p_2^1 and p_3^1 for simplicity. The symbol ℓ in these equations represents a normalized Lagrange interpolating function, as given in Equations (4.4.4-6) through (4.4.4-10). Recall that the integers i, j, and k in Equation (6.4.3-1) represent the indices that identify the node point within the triangle at which $\tau_{ijk}^{(M)}$ equals unity.

This method for calculating Γ for $\tau_{ijk}^{(M)}$ and $\tau_{mno}^{(M)}$ is now outlined. From Equation (6.3.1-3), we have

$$\Gamma(\tau_{ijk}^{(M)}, \tau_{mno}^{(M)}) = \int_{A_\Gamma} \nabla\tau_{ijk}^{(M)} \cdot \nabla\tau_{mno}^{(M)}\, da \qquad (6.3.3\text{-}2)$$

From Equation (6.3.3-1), we have

$$\nabla\tau_{ijk}^{(M)} = \sum_{c=1}^{3} \left(\frac{\partial\tau_{ijk}^{(M)}}{\partial\Lambda_c}\right) \nabla\Lambda_c \qquad (6.3.3\text{-}3)$$

and

$$\nabla\tau_{mno}^{(M)} = \sum_{q=1}^{3} \left(\frac{\partial\tau_{mno}^{(M)}}{\partial\Lambda_q}\right) \nabla\Lambda_q \qquad (6.3.3\text{-}4)$$

where c and q are integer indices.

From Equations (6.3.3-2), (6.3.3-3), and (6.3.3-4), we have

$$\Gamma(\tau_{ijk}^{(M)}, \tau_{mno}^{(M)}) = \sum_{c=1}^{3} \sum_{q=1}^{3} \int_{A_\Gamma} \left(\frac{\partial\tau_{ijk}^{(M)}}{\partial\Lambda_c}\right)\left(\frac{\partial\tau_{mno}^{(M)}}{\partial\Lambda_q}\right)(\nabla\Lambda_c \cdot \nabla\Lambda_q)\, da \qquad (6.3.3\text{-}5)$$

where A is the area of the triangle. As we saw in Section 6.3.2, the dot product

$$\nabla\Lambda_c \cdot \nabla\Lambda_q$$

is constant over A, and furthermore,

$$\nabla\Lambda_c \cdot \nabla\Lambda_q = \frac{1}{A}\Gamma(\Lambda_c, \Lambda_q) = \frac{1}{A}\Gamma(p_c^1, p_q^1) \qquad (6.3.3\text{-}6)$$

where the latter are evaluated in Equations (6.3.2-2) through (6.3.2-7). Then, from Equation (6.3.3-5),

$$\Gamma(\tau_{ijk}^{(M)}, \tau_{mno}^{(M)}) = \sum_{c=1}^{3} \sum_{q=1}^{3} \nabla\Lambda_c \cdot \nabla\Lambda_q \int_{A} \left(\frac{\partial\tau_{ijk}^{(M)}}{\partial\Lambda_c}\right)\left(\frac{\partial\tau_{mno}^{(M)}}{\partial\Lambda_q}\right) da \qquad (6.3.3\text{-}7)$$

From Equations (4.4.4-8) and (6.3.3-1), we can see that

$$\frac{\partial \tau_{ijk}^{(M)}}{\partial \Lambda_c}, \qquad 1 \leqslant c \leqslant 3$$

can be expressed as a polynomial in Λ_1, Λ_2, and Λ_3. The same is true for

$$\frac{\partial \tau_{mno}^{(M)}}{\partial \Lambda_q}, \qquad 1 \leqslant q \leqslant 3$$

and for the integrand in Equation (6.3.3-7). That is, this integrand can be expressed as the linear combination of, say, N_T polynomial terms, as

$$\sum_{u=1}^{N_T} a_u \Lambda_1^{p(u)} \Lambda_2^{w(u)} \Lambda_3^{r(u)}$$

where $p(u)$, $w(u)$, and $r(u)$ are integer exponents. Equation (6.3.3-7) becomes

$$\Gamma(\tau_{ijk}^{(M)}, \tau_{mno}^{(M)}) = \sum_{c=1}^{3} \sum_{q=1}^{3} \nabla\Lambda_c \cdot \nabla\Lambda_q \sum_{u=1}^{N_T} \int_A \Lambda_1^{p(u)} \Lambda_2^{w(u)} \Lambda_3^{r(u)}\, da \qquad (6.3.3\text{-}8)$$

Silvester (Ref. 6, pp. 95–100, eq. 15) shows that for a triangle,

$$\int_A \Lambda_1^{p(u)} \Lambda_2^{w(u)} \Lambda_3^{r(u)}\, da = A \frac{p(u)!\; w(u)!\; r(u)!}{[p(u) + w(u) + r(u) + 2]} \qquad (6.3.3\text{-}9)$$

so that Equation (6.3.3-8) becomes

$$\Gamma(\tau_{ijk}^{(M)}, \tau_{mno}^{(M)}) = \sum_{c=1}^{3} \sum_{q=1}^{3} \Gamma(\Lambda_c, \Lambda_q) \sum_{u=1}^{N_T} a_u \frac{p(u)!\; w(u)!\; r(u)!}{[p(u) + w(u) + r(u) + 2]!} \qquad (6.3.3\text{-}10)$$

6.3.4. Computation of Γ for a Shape Function over a Rectangular Finite Element

The computation of Γ for a higher order shape function over a rectangle is outlined below. The case for the lowest order shape function is then a simple example of this. Again, we start from Equation (6.3.2-1) and use Equation (4.4.6-2), so that

$$\Gamma(p_{ij}, p_{op}) = \int_{y=0}^{y_n} \int_{x=0}^{x_m} \left[L_j^n(y) L_p^n(y) \frac{dL_i^m(x)}{dx} \frac{dL_o^m(x)}{dx} \right.$$

$$\left. + L_i^m(x) L_o^m(x) \frac{dL_j^n(y)}{dy} \frac{dL_p^n(y)}{dy} \right] dx\, dy \qquad (6.3.4\text{-}1)$$

where A_Γ is, in this case, the area of the rectangle in Figure 4-8 and L represents a Lagrange interpolating function as defined in Equation (4.4.4-1). From Equation (6.3.4-1), we can see that for any order this integral is, in concept, easy to evaluate—just tedious.*

6.4. SAMPLE PROBLEM

A rather simple sample problem is worked out in this section in order to give the reader a feeling for how the information presented in Chapters 2, 3, 4, and 5, as well as in Sections 6.2 and 6.3, fit together to form the basis for working out a finite element problem.

6.4.1. Problem Definition

A two-dimensional interior problem is chosen. As shown in Section 6.2, the finite element method is best suited to an interior problem since its problem domain is bounded, with boundary conditions applied over its boundaries. Chapter 7 deals with extension of the finite element method to exterior problems. But, as we will see, this extension to exterior problems is quite complicated and is therefore unsuited to the objectives of a sample problem.

Figure 6-3 shows the configuration. Basically, we have a permanent magnet in the shape of a bar that is surrounded by free space, which, in turn, is surrounded by a rectangular picture frame of permeable material. Both the bar magnet and the picture frame are assumed to be infinitely permeable. That being the case, we can establish boundary conditions over the exterior of the bar magnet and the interior of the picture frame. Then the problem domain (in this case the *field* problem domain) is just the free space outside the bar magnet and inside the permeable picture frame. Since no current flows in the problem domain, we have, from Equation (2.2.2-1), that the curl of H equals zero, so that we can represent H by

$$H = -\nabla\phi_m \tag{6.4.1-1}$$

where ϕ_m is the reduced scalar potential. Furthermore, in this domain we have, from Equation (2.2.1-3), that

$$0 = \nabla \cdot B = \nabla \cdot \mu_0 H = \nabla \cdot H$$

with the result that throughout the problem domain,

---

*This tedium of the integration has sometimes been avoided by making a Gauss-quadruture numerical integration.

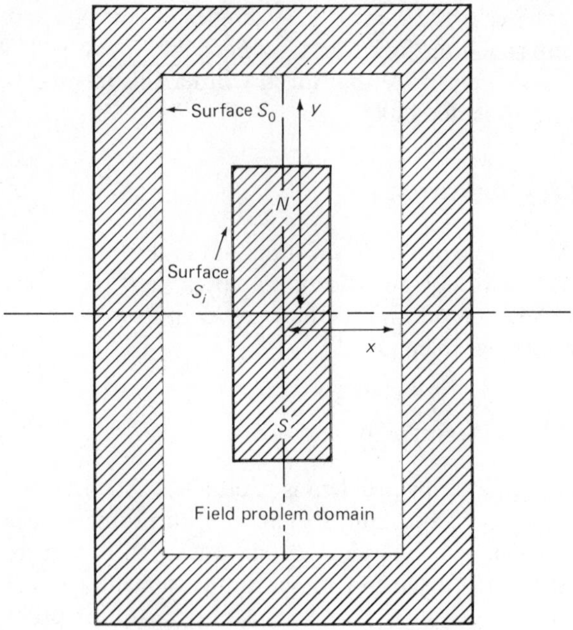

Fig. 6-3. Configuration for sample problem.

$$\nabla^2 \phi_m = 0 \tag{6.4.1-2}$$

that is, Laplace's equation holds.

We have boundary conditions to establish both on the exterior surface of the bar magnet, S_i, and the interior surface of the picture frame, S_0, as shown in Figure 6.3. Since the picture frame has infinite permeability, the component of magnetic field tangential to the picture frame is zero. Then, from Equation (6.4.1-1), we see that ϕ_m is constant over the picture frame. We take this constant to be zero; that is, $\phi_m = 0$ over the picture frame. We have established a Dirichlet boundary condition over S_0.

We know the bar magnet contains a pattern of magnetization that, in turn, produces a demagnetizing magnetic field. [See Section 2.5.3.] The sum of magnetization **M** and the demagnetization field, $\mathbf{H}_d$, equals the magnetic flux density, **B**. **B** in turn has a nonzero component normal to S_i, B_n, that we assume known. From this the Neumann boundary condition for any point P on S_i is

$$B_n(P) = -\mu_0 \frac{\partial \phi_m}{\partial n}(P) \tag{6.4.1-3}$$

In the bar magnet, the magnetic charge density [which, by Equation (2.5.3-9), equals the divergence of **M**], is concentrated near its ends. That is, there is a positive magnetic charge density near one end of the bar magnet and a negative charge density near the other end. This being the case, to a good enough approximation for this problem, we say that

$$\frac{\partial \phi_m}{\partial n} = y = h \qquad (6.4.1\text{-}4)$$

where y is the spatial coordinate along the bar magnet, as shown in Figure 6-3.

6.4.2. Construction of Hilbert Space for Problem

As discussed above, the Hilbert space, $\mathscr{H}$, comprises all functions defined over the problem domain that satisfy the essential boundary condition. In this case, the essential boundary condition is the Dirichlet boundary condition over the surface S_0 in Figure 6-3. Accordingly, each element of $\mathscr{H}$ is a function defined over the problem domain (defined in the section above) that satisfies the Dirichlet boundary condition over S_0. Since this problem does not use a mixed boundary condition [given in Equation (3.7.1-3)],

$$\sigma = 0 \qquad (6.4.2\text{-}1)$$

Then from Equation (6.2.2-3), the inner product between two elements **u** and **w** of $\mathscr{H}$ is

$$\langle \mathbf{u}, \mathbf{w} \rangle = \int_D (\nu \nabla \mathbf{u}) \cdot \nabla \mathbf{w} \, dS \qquad (6.4.2\text{-}2)$$

Next, we must construct $\mathscr{H}_N$, the N-dimensional subspace of $\mathscr{H}$. To do this, we first subdivide the problem domain into triangular elements, as shown in Figure 6-4. For simplicity, we use linear shape functions over these triangular elements. [As discussed in the preceding section, we must use pyramid-type basis functions. Figure 4-1b, then, shows what one of our basis functions looks like.] Since we use linear shape functions, we have node points only at the vertices of the triangles.

6.4.3. Construction of Linear System for Sample Problem

In this section we simply adapt Equation (6.2.3-4) for the linear system to the problem at hand. Since this problem has no mixed boundary conditions,

Equation (6.4.2-1) holds, and, from Equation (6.3.1-4) the elements of the matrix, M, are given by

$$m_{ij} = \sum_{k=1}^{n(i,j)} v_k \Gamma_k(p_i, p_j) \tag{6.4.3-1}$$

In this equation, i and j are global indices of basis functions $\boldsymbol{\alpha}_i$ and $\boldsymbol{\alpha}_j$ (and their node points). The number of finite elements in the intersection of the supports of these two basis functions in $n(i,j)$, and k is a local index used to designate each of these finite elements.

The symbol v_k in Equation (6.4.3-1) can represent relative permeability of each finite element in the intersection of supports. (If this were a static electric problem, then v_k could represent the permittivity of each of these finite elements.) That is, if we have a problem in which the medium differs for different finite elements, then these differences are represented by different values of v_k. In this problem, the entire problem domain is free space, with the result that v_k equals unity for all finite elements, and Equation (6.4.3-1) becomes

$$m_{ij} = \sum_{k=1}^{n(i,j)} \Gamma_k(p_i, p_j) = \langle \boldsymbol{\alpha}_j, \boldsymbol{\alpha}_i \rangle_2 \tag{6.4.3-2}$$

The elements m_{ij} of the matrix M are given by Equation (6.4.3-2). Expressions for $\Gamma_k(p_i, p_j)$ in this equation are given by Equations (6.3.2-2) through (6.3.2-7). [Notice that the indices i and j in Equation (6.4.3-2) are the global indices of those node points. The indices in Equations (6.3.2-2) through (6.3.2-7) are the local indices 1, 2, and 3 of the vertices of the triangular finite element. That is, each time we apply one of these equations to a triangle, we use both a global index and a local index at each of its vertices.]

Finally, we use Equation (6.2.3-3) to compute the terms b_i that are used in Equation (6.2.3-4). In this application, since we do not use a mixed boundary condition, σ equals zero. Since our magnetic potential ϕ_m obeys Laplace's equation (Equation (6.4.1-2), then f [as given in Equation (6.2.1-8)] equals zero. Furthermore, since a Dirichlet boundary condition of zero is required over S_0, then $\boldsymbol{\beta}_b$ [in Equation (6.2.1-4)] equals zero. From this fact, and Equation (6.2.1-7), we have

$$\beta_i = 0, \qquad 1 + N_a \leqslant i \leqslant N \tag{6.4.3-3}$$

With these restrictions, Equation (6.2.3-3) becomes

$$b_i = \int_{\partial N} \boldsymbol{\alpha}_i h \, dS \tag{6.4.3-4}$$

It remains, then, to develop an algorithm for evaluating b_i for $1 \leqslant i \leqslant N_a$ from Equation (6.4.3-4). First, notice that ∂N is that portion of the boundary over which the natural boundary condition, or in this case, the Neumann boundary condition, is applied. For this problem, then, ∂N is the same as the exterior surface of the bar magnet, S_i. Furthermore, this boundary condition is given in Equation (6.4.1-4). Then, from that equation and Equation (6.4.3-4),

$$b_i = \int_{S_i} \alpha_i y \, dS \tag{6.4.3-5}$$

Notice that b_i is given by Equation (6.4.3-5) only if the basis function α_i intersects the boundary surface S_i. Otherwise, b_i equals zero. Again, to evaluate this integral, we split α_i up into the shape functions over its supports. That is, Equation (6.4.3-5) becomes

$$b_i = \sum_{k=1}^{n(i)} \int_{S_{ik}} p_{ik}^1(x, y) y \, dS, \qquad 1 \leqslant i \leqslant N_a \tag{6.4.3-6}$$

In this equation, $n(i)$ is the number of finite elements in the support of basis function i that intersect S_i, and k is a local index of these finite elements. Further, $p_{ik}(x, y)$ is the linear shape function over finite element k that equals unity at node point i, and S_{ik} is the intersection of $p_{ik}(x, y)$ and S_i. Since both $p_{ik}(x, y)$ and y vary linearly over S_{ik}, the integrals in this equation are simply evaluated analytically.

Equations (6.4.3-2) and (6.4.3-6) were used to compute the linear system that was, in turn, solved to compute the field values presented in the next section.

6.4.4. Computed Results for Sample Problem

Figure 6-4 shows the computed results for this sample problem. The solid curves are contours of constant reduced magnetic scalar potential, taken at equally spaced levels. The dashed lines are, again, the boundaries of the triangular finite elements that are used. In this computation, a total of 198 node points are used. Of these, 60 are along the picture frame and a field value of zero is assigned at these node points at the outset of the problem. The field was calculated at the remaining 138 points by solving the linear system. A total of 304 finite elements was used.

Notice that this configuration and its field are symmetric about both the x and y axes. The field could have been found, then, by computing it only in

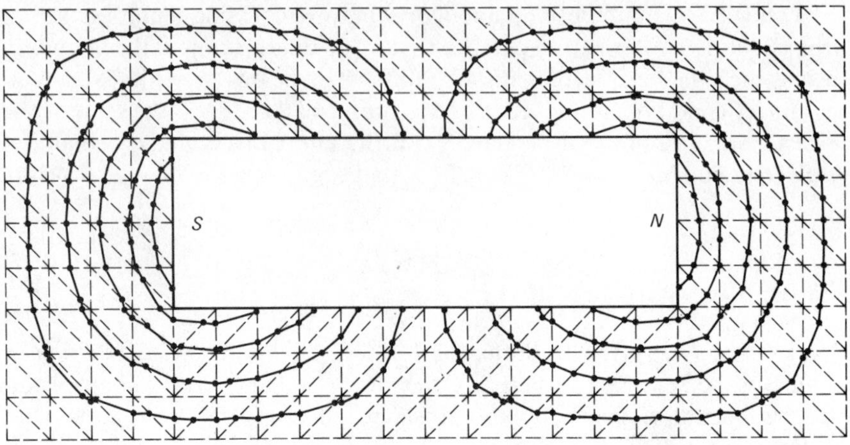

Fig. 6-4. Finite elements and computed results for sample problem.

one of these quadrants. This would have been more efficient, by way of saving in both computer time and computer memory.

REFERENCES

1. Mitchel, A. R., and Wait, R.. *The Finite Element Method in Partial Differential Equations.* New York: John Wiley & Sons, 1977.
2. Zienkiewitz, O. C.. *The Finite Element Method in Engineering Science.* New York, McGraw-Hill Book Company, 1971.
3. Strang, G., and Fix, G. J.. *An Analysis of the Finite Element Method.* Englewood Cliffs, N.J.: Prentice-Hall, Inc., 1973.
4. Wexler, Alvin. *Finite Elements for Technologists.* Winnipeg: Electrical Engineering Department, University of Manitoba, 1974.
5. Chari, M. V. K., and Silvester, P. P., Editors. *Finite Elements in Electrical and Magnetic Field Problems.* New York: John Wiley & Sons, 1980.
6. Silvester, P. Symmetric Quadrature Formulae for Simplexes, Mathematics of Computation, January 1970, Vol. 24, No. 109.

7
FINITE ELEMENT METHOD FOR EXTERIOR PROBLEMS

7.1. INTRODUCTION

The exterior problem for which solutions are presented in this chapter is depicted in Figure 7-1. As shown, all medium discontinuities and sources are confined within some closed surface, S_0. This figure also shows a region, R_e, inside S_0 that is excluded from the problem domain. The problem includes suitable boundary conditions on ∂R_e, the boundary of R_e. The algorithms presented below can be adapted to one or many such excluded regions, or none at all.

The solutions of the exterior problem presented here can also be used as the *scattered* field in problems that differ from these only by the addition of sources at infinity.

As shown in Chapter 6, the finite element method is readily adapted to interior problems. To adapt the method to exterior problems, we augment the algorithm that we use in interior problems. This augmentation can be accomplished in a variety of ways, by both iterative and simultaneous methods.

Using an iterative method, a sequence of fields is computed, each field being more accurate than the last. An algorithm has been reported (Ref. 1) in which the *finite difference* solution to an interior problem was augmented to the solution of an exterior problem. However, the method of augmentation used could be applied to the finite element method as well.

Using a simultaneous method, the augmentation of the interior problem algorithm results in a set of simultaneous equations that can be solved for the field. This field is the solution to the exterior problem.

Simultaneous methods have the advantage over iterative methods in that the field is computed only once rather than several times. For this reason, the presentation below is restricted to two simultaneous methods, one by McDonald and Wexler (Ref. 2) and the other by Silvester et al. (Ref. 3).

7.2. MCDONALD-WEXLER ALGORITHM

Consider the configuration for the problem shown in Figure 7-1. At the onset of the problem, we know neither the potential, γ, nor its normal derivative $\partial\gamma/\partial n$ over the surface, S_0. However, we do know that, outside S_0, there is only free space. And that very fact provides the key to this algorithm.

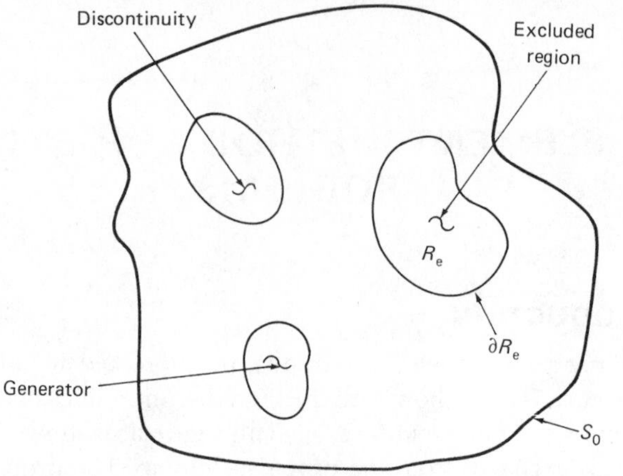

Fig. 7-1. General configuration for exterior problem.

As shown below and in Appendix C, this fact, along with Green's theorem, enables us to express the normal derivative $\partial\gamma/\partial n$ over S_0 as a linear transform of the values of γ over S_0. This linear transform, then, expresses the effect of the free space *outside* S_0 upon the field. In addition, we use certain equations developed in Chapter 6 to express the fields *inside* S_0. These equations for the fields inside S_0 are all that we need for the McDonald-Wexler algorithm.

The steps involved in this algorithm are the following:

1. For this problem, construct the structure of node points and finite elements over the problem domain, as shown in Figure 7-2.
2. Formulate the field inside and on S_0 in terms of its derivative normal to S_0.
3. Formulate the derivative of the field normal to S_0 in terms of the field on S_0.
4. Combine these formulations to provide a system of equations to be solved for the field, and solve this system.

Sections 7.2.1, 7.2.2, 7.2.3, and 7.2.4 discuss items 1, 2, 3, and 4, respectively.

7.2.1. Structure of Node Points and Finite Elements

Figure 7-2 shows the structure of node points and finite elements used in this algorithm. There are N_d essential, Dirichlet node points deployed along ∂E (that portion of the boundary of the excluded region, R_e, over which a Dirichlet boundary condition is employed). All remaining node points are

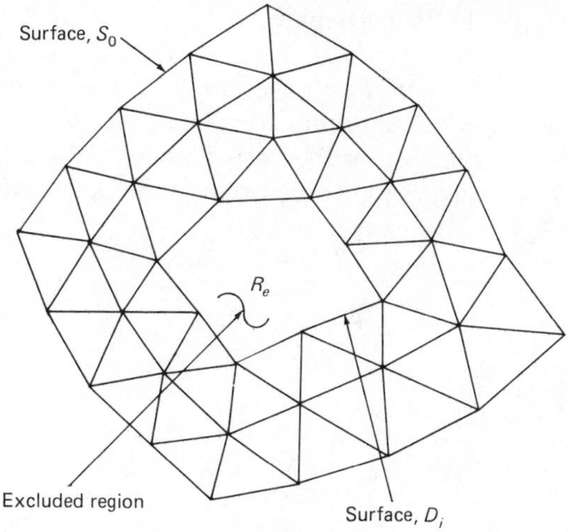

Fig. 7-2. Configuration for McDonald-Wexler algorithm.

active node points. Letting N_a be the number of active node points and N the total number of node points, we have

$$N_a + N_d = N$$

Notice that a boundary condition is not imposed over S_0 (the outer boundary of the domain, D). As discussed above, both the field *and* its normal derivatives over S_0 are computed (approximately) simultaneously in this algorithm.

The finite elements used in Figure 7-2 and the shape functions and basis functions defined over them are constructed as discussed in Chapter 4.

As in Chapter 6, we have a Hilbert space, $\mathscr{H}$, whose elements are all continuous functions defined over the domain, D, of this problem. The space $\mathscr{H}$ has the same inner product as used in Chapter 6. Again, we have an exact solution, $\boldsymbol{\gamma}$, and an approximate finite element solution $\boldsymbol{\beta}$ that are elements of $\mathscr{H}$. Furthermore, $\boldsymbol{\beta}$ is given by

$$\boldsymbol{\beta} = \sum_{i=1}^{N} \beta_i \boldsymbol{\alpha}_i \tag{7.2.1-1}$$

where the $\boldsymbol{\alpha}_i$ are basis functions constructed from the node points, finite elements, and shape functions as shown in Chapter 4. Then these basis functions

$$\boldsymbol{\alpha}_i, \qquad 1 \leqslant i \leqslant N$$

constitute a basis for the Hilbert space, $\mathscr{H}_N$, which is a finite-dimensional subspace of $\mathscr{H}$.

As in Chapters 5 and 6, $\boldsymbol{\beta}$ is partitioned into $\boldsymbol{\beta}_h$ and $\boldsymbol{\beta}_b$, where $\boldsymbol{\beta}_h$ is a linear combination of the basis functions that correspond to the active node points, and $\boldsymbol{\beta}_b$ is a linear combination of the basis functions of the Dirichlet node points. If, for convenience, we give the active node points the indices i for $1 \leqslant i \leqslant N_a$, then $\boldsymbol{\beta}_h$ and $\boldsymbol{\beta}_b$ are expressed by

$$\boldsymbol{\beta}_h = \sum_{i=1}^{N_a} \beta_i \boldsymbol{\alpha}_i \qquad (7.2.1\text{-}2)$$

and

$$\boldsymbol{\beta}_b = \sum_{i=N_a+1}^{N} \beta_i \boldsymbol{\alpha}_i \qquad (7.2.1\text{-}3)$$

Notice that the β_i that correspond to the Dirichlet node points are fixed by the Dirichlet boundary conditions. Therefore, $\boldsymbol{\beta}_b$ is known once the node points, finite elements, shape functions, and basis functions are defined. Alternatively, we must solve the problem for $\boldsymbol{\beta}_h$.

7.2.2. Formulation of the Field Inside S_0

In this problem we require that γ satisfy the equation

$$-\nabla \cdot k\nabla\gamma = f \qquad (7.2.2\text{-}1)$$

inside domain D (excluding Re) and

$$\nabla^2\gamma = 0 \qquad (7.2.2\text{-}2)$$

outside S_0. Furthermore γ and its normal derivatives are continuous over ∂D.

The derivation starts off similar to the derivation given in Chapter 6 for the formulation of the solution of the interior problem by the finite element method. Using Equation (7.2.2-1), we have

$$\langle -\nabla \cdot k\nabla\gamma - f, \boldsymbol{\alpha}_i \rangle = \int_D (-\nabla \cdot k\nabla\gamma - f)\boldsymbol{\alpha}_i \, dv = 0, \quad 1 \leqslant i \leqslant N_a \qquad (7.2.2\text{-}3)$$

where $\boldsymbol{\alpha}_i$ is the basis function that corresponds to an active node point. We apply an integration by parts to the first term in the bracket of the integrand of this equation and obtain

$$\int_D \nabla \boldsymbol{\alpha}_i \cdot k\gamma \, dv = \int_D f\boldsymbol{\alpha}_i \, dv + \int_{\partial D_i} \boldsymbol{\alpha}_i \mathbf{n} \cdot k\nabla\gamma \, dS + \int_{S_0} \boldsymbol{\alpha}_i \mathbf{n} \cdot k\nabla\gamma \, dS \qquad (7.2.2\text{-}4)$$

In this equation, ∂D_i is the boundary surface of the excluded region, R_e, and S_0 is the outer boundary of domain D, as shown in Figures 7-1 and 7-2. Now ∂D_i consists entirely of just two parts, ∂E and ∂N. Furthermore, $\boldsymbol{\alpha}_i$ (being the basis function of an active node point) equals zero over ∂E. Therefore, the domain of integration of the first surface integral of Equation (7.2.2-4) can be reduced from ∂D_i to ∂N. Furthermore, over ∂N, by definition, a Neumann boundary condition,

$$h = \mathbf{n} \cdot k\nabla\gamma$$

is imposed (as discussed in Chapter 3). Then we have

$$\int_{\partial D_i} \boldsymbol{\alpha}\mathbf{n} \cdot k\nabla\gamma \, dS = \int_{\partial N} \boldsymbol{\alpha}_i \mathbf{n} \cdot k\nabla\gamma \, dS = \int_{\partial N} \boldsymbol{\alpha}_i h \, dS$$

When this equation is combined with Equation (7.2.2-4), we can write

$$\int_D \nabla \boldsymbol{\alpha}_i \cdot k\nabla\gamma \, dv - \int_{S_0} \boldsymbol{\alpha}_i \mathbf{n} \cdot k\nabla\gamma \, dS = C(\boldsymbol{\alpha}_i), \qquad 1 \leqslant i \leqslant N_a \qquad (7.2.2\text{-}5)$$

where

$$C(\boldsymbol{\alpha}_i) = \int_D f\boldsymbol{\alpha}_i \, dv + \int_{\partial N} \boldsymbol{\alpha}_i h \, dS \qquad (7.2.2\text{-}6)$$

Since f is known over D, and h is known over ∂N, we see in this equation that C is, in fact, a linear functional. Putting it another way, for any given $\boldsymbol{\alpha}_i$, $C(\boldsymbol{\alpha}_i)$ is known. The only significant difference between these two equations and Equation (6.2.1-14) of Chapter 6 is the inclusion of the integral over S_0 in Equation (7.2.2-5).

In this algorithm we compute two approximating functions. The first of these is the desired scalar field $\boldsymbol{\beta}$ (discussed on Section 7.2.1), which approximates γ, as discussed in Chapter 6. In addition, the second function, $\boldsymbol{\lambda}$, approximates

$$-\mathbf{n} \cdot k\nabla\gamma$$

over S_0. Then, we require that $\boldsymbol{\beta}$ and $\boldsymbol{\lambda}$ satisfy Equation (7.2.2-5), so that

$$\int_D \nabla \boldsymbol{\alpha}_i \cdot k \nabla \boldsymbol{\beta} \, dv + \int_{S_0} \alpha_i \boldsymbol{\lambda} \, dS = C(\boldsymbol{\alpha}_i) \tag{7.2.2-7}$$

To obtain a numerical solution, both $\boldsymbol{\beta}$ and $\boldsymbol{\lambda}$ must be expressed as linear combinations of certain basis functions. This was done for $\boldsymbol{\beta}$ in the previous section. We express $\boldsymbol{\lambda}$ by

$$\boldsymbol{\lambda}(P_0) = \sum_{j=1}^{N_a} \lambda_j \boldsymbol{\alpha}_j(P_0) \tag{7.2.2-8}$$

In this equation, P_0 is a point on S_0. That is, $\boldsymbol{\alpha}_j(P_0)$ is defined along the intersection of S_0 and the support of $\boldsymbol{\alpha}_j$. Notice that for most of the node points, j, their basis functions, $\boldsymbol{\alpha}_j$, have supports that do not intersect S_0. For these basis functions,

$$\boldsymbol{\alpha}_j(P_0) = 0$$

We proceed now to introduce into Equation (7.2.2-7) the expansions of $\boldsymbol{\beta}$ and $\boldsymbol{\lambda}$ in terms of basis functions that have been developed above. But first, it is helpful to partition $\boldsymbol{\beta}$ into $\boldsymbol{\beta}_h$ and $\boldsymbol{\beta}_b$, since, as discussed above, $\boldsymbol{\beta}_b$ is determined at the outset by the boundary conditions imposed at the Dirichlet node points, while $\boldsymbol{\beta}_h$ is to be determined by the solution of the problem. Then, Equation (7.2.2-7) becomes

$$\int_D \nabla \boldsymbol{\alpha}_i \cdot k \nabla \boldsymbol{\beta}_h \, dv + \int_{S_0} \boldsymbol{\alpha}_i \boldsymbol{\lambda} \, dS = C(\boldsymbol{\alpha}_i) - \int_D \nabla \boldsymbol{\alpha}_i \cdot k \boldsymbol{\beta}_b \, dv, \qquad 1 \leqslant i \leqslant N_a$$

and with Equations (7.2.1-2), (7.2.1-3), and (7.2.2-8), we have

$$\sum_{j=1}^{N_a} \beta_i \int_D \nabla \boldsymbol{\alpha}_i \cdot k \nabla \boldsymbol{\alpha}_j \, dv + \sum_{j=1}^{N_a} \lambda_j \int_{\partial S_0} \boldsymbol{\alpha}_i \boldsymbol{\alpha}_j \, dS$$

$$= C(\boldsymbol{\alpha}_i) - \sum_{j=N_a+1}^{N_T} \beta_i \int_D \nabla \boldsymbol{\alpha}_i \cdot k \nabla \boldsymbol{\alpha}_j \, dv, \qquad 1 \leqslant i \leqslant N_a \tag{7.2.2-9}$$

Notice that the left sides of the above two equations are to be determined by the solution to the problem, while the right sides are known at the outset.

For computational purposes it is convenient to put Equation (7.2.2-9) in the matrix-vector form:

$$S\hat{\boldsymbol{\beta}} + G\hat{\boldsymbol{\lambda}} = \mathbf{b} \tag{7.2.2-10}$$

Comparison of Equations (7.2.2-9) and (7.2.2-10) shows that S and G are both square matrices of order N_a that have, as elements,

$$s_{ij} = \int_D \nabla\alpha_i \cdot k\nabla\alpha_j \, dv \tag{7.2.2-11}$$

and

$$g_{ij} = \int_{\partial S_0} \alpha_i \alpha_j \, dS, \qquad 1 \leqslant i, j \leqslant N_a \tag{7.2.2-12}$$

Furthermore, $\hat{\boldsymbol{\beta}}$ and $\hat{\boldsymbol{\lambda}}$ are both vectors of N_a elements each, having, respectively, the elements β_i and λ_i for $1 \leqslant i \leqslant N_a$. The vector $\mathbf{b}$ has elements

$$b_i = C(\alpha_i) - \sum_{j=N_a+1}^{N} \int_D \nabla\alpha_i \cdot k\nabla\alpha_j \, dv, \qquad 1 \leqslant i \leqslant N_a \tag{7.2.2-13}$$

7.2.3. Formulation for the Derivative of Field Normal to S_0 in Terms of the Field on S_0

The region outside S_0 has been assumed to be free space, and free of all sources (including those at infinity). Thus, the potential, γ, obeys Laplace's equation (as discussed above). Appendix C shows that for this situation, if the field point, $\mathbf{r}_f$, is taken *on* S_0, then we have the equation

$$\tfrac{1}{2}\gamma(\mathbf{r}_f) = \int_{S_0} \gamma(\mathbf{r}_s) \frac{\partial G_3}{\partial n}(\mathbf{r}_s, \mathbf{r}_f) \, dS_s - \int_{S_0} G_3(\mathbf{r}_s, \mathbf{r}_f) \frac{\partial \gamma(\mathbf{r}_s)}{\partial n} \, dS_s \tag{7.2.3-1}$$

In this equation, the normal derivatives are taken in the direction *outward* from S_0, and the integrations over S_0 are taken by varying the source point, $\mathbf{r}_s$. The Green's function of free space, G_3, is given by

$$G_3(\mathbf{r}_s, \mathbf{r}_f) = \frac{1}{4\pi |\mathbf{r}_s - \mathbf{r}_f|}$$

For problems in two dimensions, this is replaced by the Green's Function of free space in two dimensions, G_2, given by

$$G_2(\mathbf{r}_s, \mathbf{r}_f) = \frac{-1}{2\pi} \ln |\mathbf{r}_s - \mathbf{r}_f|$$

If Equation (7.2.3-1) is multiplied by $\boldsymbol{\alpha}_i$ and integrated over S_0 by varying the field point, $\mathbf{r}_f$, we obtain

$$\int_{S_0} \boldsymbol{\alpha}_i(\mathbf{r}_f)\left[\tfrac{1}{2}\gamma(\mathbf{r}_f) - \int_{S_0} \gamma(\mathbf{r}_s)\frac{\partial}{\partial n} G_3(\mathbf{r}_s, \mathbf{r}_f)\, dS_s + \int_{S_0} G_3(\mathbf{r}_s, \mathbf{r}_f)\frac{\partial}{\partial n}\gamma(\mathbf{r}_s)\, dS_s\right] dS_f = 0 \qquad (7.2.3\text{-}2)$$

As in the previous section, we approximate γ and its normal derivative, in this equation by $\boldsymbol{\beta}_h$ and $\boldsymbol{\lambda}$. When this is done, and the expansions for $\boldsymbol{\beta}_h$ and $\boldsymbol{\lambda}$ given in Equations (7.2.1-2) and (7.2.2-8) are used, we have

$$\sum_{j=1}^{N_a} \beta_j \int_{S_0} \boldsymbol{\alpha}_i(\mathbf{r}_f)\left[\tfrac{1}{2}\boldsymbol{\alpha}_j(\mathbf{r}_f) - \int_{S_0} \boldsymbol{\alpha}_j(\mathbf{r}_s)\frac{\partial}{\partial n} G(\mathbf{r}_s, \mathbf{r}_f)\, dS\right] dS_f$$
$$- \sum_{j=1}^{N_a} \lambda_j \int_{S_0} \boldsymbol{\alpha}_i(\mathbf{r}_f) \int_{S_0} G_3(\mathbf{r}_s, \mathbf{r}_f)\frac{\partial}{\partial n}\boldsymbol{\alpha}_j(\mathbf{r}_s)\, dS_s\, dS_f = 0, \qquad 1 \leqslant i \leqslant N_a \qquad (7.2.3\text{-}3)$$

Again, this equation is put into matrix-vector form:

$$C\hat{\boldsymbol{\beta}} - Z\hat{\boldsymbol{\lambda}} = \mathbf{0} \qquad (7.2.3\text{-}4)$$

where C and Z are both square matrices of order N_a that have, as elements,

$$c_{ij} = \int_{S_0} \boldsymbol{\alpha}_i(\mathbf{r}_f)\left[\tfrac{1}{2}\boldsymbol{\alpha}_j(\mathbf{r}_f) - \int_{S_0} \boldsymbol{\alpha}_j(\mathbf{r}_s)\frac{\partial}{\partial n} G(\mathbf{r}_s, \mathbf{r}_f)\, dS_s\right] dS_f \qquad (7.2.3\text{-}5)$$

and

$$z_{ij} \times \int_{S_0} \boldsymbol{\alpha}_i(\mathbf{r}_f) \int_{S_0} G_3(\mathbf{r}_s, \mathbf{r}_f)\frac{\partial}{\partial n}\boldsymbol{\alpha}_j(\mathbf{r}_s)\, dS_s\, dS_f \qquad (7.2.3\text{-}6)$$

and $\hat{\boldsymbol{\beta}}$ and $\hat{\boldsymbol{\lambda}}$ have the definitions given above.

7.2.4. Linear System for Solution

The formulations derived in the last two subsections can be combined together to compute $\hat{\boldsymbol{\beta}}$ and $\hat{\boldsymbol{\lambda}}$. Using Equations (7.2.2-10) and (7.2.3-4), we have the matrix-vector system

$$\begin{bmatrix} S & G \\ C & -Z \end{bmatrix} \begin{bmatrix} \hat{\boldsymbol{\beta}} \\ \hat{\boldsymbol{\lambda}} \end{bmatrix} = \begin{bmatrix} \mathbf{b} \\ \mathbf{0} \end{bmatrix} \tag{7.2.4-1}$$

The first pair of brackets in this equation contains a square matrix, of order $2N_a$, of which S, G, C, and Z are submatrices. The second two pairs of brackets contain vectors of $2N_a$ elements each. Therefore, we have a linear system of order $2N_a$, which can be solved for $\hat{\boldsymbol{\beta}}$ and $\hat{\boldsymbol{\lambda}}$ by standard techniques. As shown above, the elements of $\hat{\boldsymbol{\beta}}$ are the coefficients in the expansion of $\boldsymbol{\beta}_h$ and are therefore desired. Alternatively, $\hat{\boldsymbol{\lambda}}$ is usually not desired, since it can be used only to approximate the normal derivative of γ over S_0.

There are certain practical simplifications that can be made in Equation (7.2.4-1). These result from fact that the elements of the submatrices G, C, and Z are all expressed as integrals over S_0. Furthermore, the values of $\boldsymbol{\alpha}_i$ and $\boldsymbol{\alpha}_j$ in these integrals are nonzero only if the supports of $\boldsymbol{\alpha}_i$ and $\boldsymbol{\alpha}_j$ intersect S_0. From this, we see that, for any i, if the support of $\boldsymbol{\alpha}_i$ does not intersect S_0, then the ith rows of these submatrices equal zero. And, for any j, if the support of $\boldsymbol{\alpha}_j$ does not intersect S_0, then the jth columns of these matrices equal zero. By eliminating the rows and columns of zeros, the order of the linear system of Equation (7.2.4-1) can be reduced from $2N_a$ to $N_a + M$ where M is the number of basis functions having supports that intersect S_0. In this case G becomes a submatrix with N_a rows and M columns, C becomes a submatrix with M rows and N_a columns, and Z becomes a submatrix with M rows and M columns. In addition, $\hat{\boldsymbol{\lambda}}$ is reduced to a vector of M elements.

Another simplification results from the elimination of $\hat{\boldsymbol{\lambda}}$ from the problem by matrix manipulation. Multiplying Equation (7.2.3-4) by Z^{-1}, we have

$$\hat{\boldsymbol{\lambda}} = Z^{-1}C\hat{\boldsymbol{\beta}} \tag{7.2.4-2}$$

Combining this with Equation (7.2.2-10) yields

$$[S + GZ^{-1}C]\hat{\boldsymbol{\beta}} = \mathbf{b} \tag{7.2.4-3}$$

This is a linear system of order N_a that can be solved directly for $\hat{\boldsymbol{\beta}}$. Notice that this equation can be used whether or not the zero rows and columns have been removed from the submatrices G, Z, and C. In the use of this equation, the removal of those zero rows and columns would only simplify computation of the matrix product $GZ^{-1}C$.

Notice that $\boldsymbol{\beta}$ over D and $\boldsymbol{\lambda}$ over S_0 satisfy Equations (7.2.2-7) and (7.2.3-2) in the same way that γ and $-\mathbf{n}\cdot K\nabla\gamma$ do. From this fact, we can show that $\boldsymbol{\beta}$ is a projection of γ onto the finite-dimensional linear space for which $\boldsymbol{\alpha}_i$ for $1 \leqslant \boldsymbol{\alpha}_i \leqslant N$ is a basis.

7.2.5. Steps of the Algorithm

This algorithm involves sufficient detail to warrant listing its steps, as given below:

1. Define the problem. If we assume that the equation to be satisfied by the field is Equation (7.2.2-1), we must define the closed surface S_0 and any excluded regions. We choose S_0 to enclose the smallest amount of space possible, such that outside S_0 there is only free space. For any excluded regions, we must define boundary conditions (either Dirichlet or Neumann or both).
2. Deploy node points and finite elements over the problem domain [the space inside S_0 and outside any excluded regions, as shown in Figure 7-1 and 7-2]. Designate which node points are Dirichlet node points and which are active node points. Assign field values to the Dirichlet node points.
3. Using the information given in Chapter 4, define shape functions over the finite elements and, from these, the pyramid-type basis functions.
4. Compute $C(\boldsymbol{\alpha}_i)$ for $1 \leqslant i \leqslant N_a$ by Equation (7.2.2-6).
5. Compute the submatrices S and G and the vector $\mathbf{b}$ by Equations (7.2.2-11), (7.2.2-12), and (7.2.2-13), respectively. In computing G, remove its zero *columns* so that it has N_a rows and M columns.
6. Compute the submatrices C and Z by Equations (7.2.3-5) and (7.2.3-6). In computing C and Z remove the zero *rows* of C and Z and the zero *columns* of Z. Then C has M rows and N_a columns, and Z has M rows and M columns.
7. Compute the vector $\hat{\boldsymbol{\beta}}$ from the matrix-vector Equation (7.2.4-3).
8. Compute $\boldsymbol{\beta}_h$ and $\boldsymbol{\beta}_b$ at any desired point in space, using Equations (7.2.1-2) and (7.2.1-3). The coefficients β_i used in Equation (7.2.1-2) are the elements of the vector $\hat{\boldsymbol{\beta}}$, and the coefficients of β_i in Equation (7.2.1-3) are the Dirichlet node point fields, evaluated in step 2 above. Then $\boldsymbol{\beta}$ is just the sum of $\boldsymbol{\beta}_h$ and $\boldsymbol{\beta}_b$.

7.3. SILVESTER ET AL. ALGORITHM

7.3.1. General Approach

This algorithm uses a structure of finite elements that extends, outside of S_0, far beyond it in every direction. The potential field is then set to zero at all of the outermost node points, and the field is computed. Since these outermost field points are so far distant from the region of interest (containing all generators), the exact field at these points is negligibly small. As a result, the error caused by setting the field to be zero at these points is negligible. The resulting field can then be a good approximation to the exact field for the exterior problem.

7.3.2. Configuration of Node Points and Finite Elements

Figure 7-3 shows the central portion of structure of node points and finite elements that is used in this algorithm. Again, all medium nonuniformities and generators are confined within S_0. Outside S_0, there is only free space. The steps used in constructing the structure in Figure 7-3 are as follows:

1. Choose a point P, roughly at the centroid of the region inside S_0.
2. Choose a number $K > 1$ (say, approximately 1.5).
3. Lay out radials that extend from point P outward through all of the node points along S_0.
4. For each radial, measure the distance, R, between P and the node point where the radial intersects S_0. Then put a node point on the radial a distance KR from P.

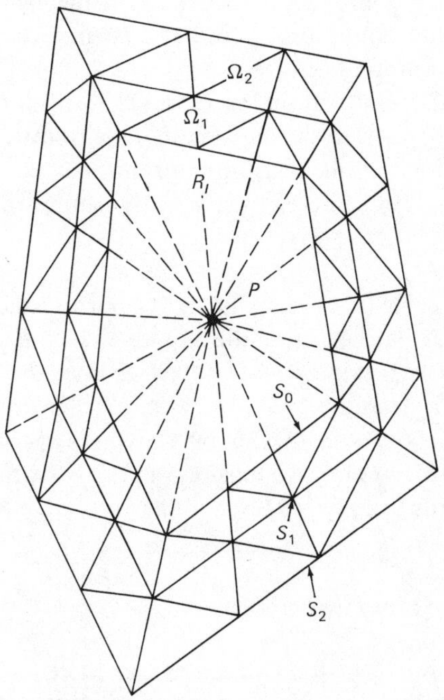

Fig. 7-3a. Configuration of exterior finite elements for Silvester et al. algorithm.

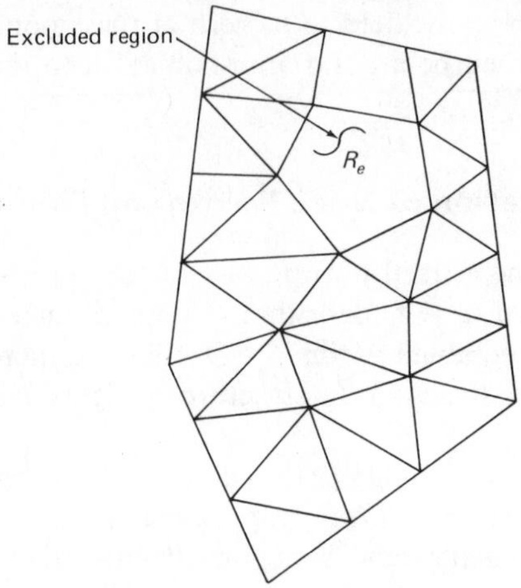

Fig. 7-3b. Configuration of region R_I with interior finite elements for Silvester et al. algorithm.

5. Establish a new surface through these new node points. The annular region between S_0 and this new surface is called Ω_1.
6. Using the node points on S_0 and the newly established node points, subdivide Ω_1 into finite elements.
7. Now put node points along each radial a distance K^2R from P and establish a new surface through these node points. The region between the previously established surface and this new surface is called Ω_2.
8. Subdivide Ω_2 into finite elements in the same way that Ω_1 was subdivided.

It is evident that each finite element in Ω_2 corresponds to just one finite element in Ω_1, and that corresponding finite elements are similar. Furthermore, all sides of any triangle in Ω_2 are larger than the corresponding sides of the coresponding triangle in Ω_1 by precisely the ratio K. Therefore, the entire region Ω_2 is similar to the region Ω_1 where *every* dimension in Ω_2 is larger than the corresponding dimension in Ω_1 by the ratio K. Thus, we call K the *mapping* factor from Ω_1 to Ω_2.

7.3.3. Hypothetical Algorithm

To understand the Silvester algorithm, it is best to first consider the following hypothetical algorithm. This hypothetical algorithm is easier to understand than the actual algorithm, but it has the same solution.

1. By the means discussed above, construct not only the annuli Ω_1 and Ω_2, but $\Omega_3, \Omega_4, \ldots, \Omega_M$, where M is some very large integer. The construction of these annuli includes the construction of the node points and finite elements within them. The union of these annuli is then a region that is bounded internally by the surface S_0 and externally by the surface S_M.
2. Construct a system of basis functions over the finite elements in these annuli. These basis functions, plus the basis functions covering the region inside S_0, constitute a basis for the Hilbert space, $\mathscr{H}_N$, of this problem.
3. Set the field at the node points of S_M to zero.
4. Construct and solve a system of linear equations for the field values at all node points inside S_M (both the node points on and inside S_0 and the node points outside S_0).

From the steps above, we see that this hypothetical algorithm is, in fact, just that of the finite element interior problem discussed in Section 6.2. The problem domain is all of space inside S_M and outside any excluded regions that might be inside S_0. A Dirichlet boundary condition of zero exists on S_M.

As shown below, it is quite practical to make the integer M [in step 1] very large. Furthermore, the distance, D_M, from point P to a node point on S_M along some radial is

$$D_M = RK^M \tag{7.3.3-2}$$

where R is the distance from P to the node point at the intersection of that radial with S_0. As shown in Chapter 3, the scalar potentials for which we solve fall off at infinity as the reciprocal of the radius squared. Thus, we can see that the exact field γ, at the node points on S_M, falls off as D_M^{-2}. Then by increasing M it is possible to make this exact field at these node points as small as desired. In turn, one can show that the field error *inside* S_M, caused by the error in the boundary condition applied at S_M, must not exceed the maximum of this boundary value error on S_M. In this way, we see that this particular error can, in a practical way, be made as small as we wish, throughout the entire problem domain.

7.3.4. Actual Algorithm

In the actual algorithm presented here, we construct and solve a linear system for *only* the field values at the node points on and inside S_0. However, the field values that this algorithm produces are exactly the same as the field values that the hypothetical algorithm would have produced at these points. That is, in going from the hypothetical algorithm to the actual algorithm, we go to an algorithm that requires very much less computation but that has the same solution. In a way, this is similar to the Gaussian elimination method of solving a system of linear equations, in which we construct a sequence of linear systems, each being easier to solve than the last, and all systems having the same solution.

There are three steps in developing the actual algorithm:

1. Formulate the relationship among the node point field values for a single annulus.
2. Eliminate from the hypothetical problem the field values at the node points outside S_0.
3. Formulate a system of equations in terms of the field values at the node points on and inside S_0 [i.e., in Region R_I of Figure 7-3].

Sections 7.3.5, 7.3.6, and 7.3.7 accomplish steps 1, 2, and 3, respectively.

7.3.5. Formulation of the Annulus Ω_1

In Figure 7-3a we see that the vertices of all the triangles in the annulus Ω_1 lie at either the surface S_0 or the surface S_1. If lowest order shape functions are used, then these vertices will comprise the only node points we have in Ω_1. However, if higher order basis functions are used, there will be additional node points midway along the sides and perhaps in the interior of these triangles. Some of these additional node points will be *outside* S_0 and *inside* S_1. The formulation of Ω_1 is then more complicated with higher order shape functions than it is with lowest order shape functions. Since the formulation for Ω_1 is straightforward in either case, it will be worked out first for higher order shape functions. After that, it will be simplified to the case of lowest order shape functions.

We use a formulation similar to that developed in Section 6.2 for the interior problem. Since there is only free space outside S_0, then, in all of the annuli, $\Omega_1, \Omega_2, \ldots, \Omega_M$, we have, in Equation (6.2.1-1),

$$\nu = 1$$

and

$$f = 0$$

As in Section 6.2, the approximate solution, $\boldsymbol{\beta}_F$, is expressed by Equation (6.2.1-4):

$$\boldsymbol{\beta}_F = \boldsymbol{\beta}_h + \boldsymbol{\beta}_b$$

Here, $\boldsymbol{\beta}_h$ is a linear combination of the basis functions of the active node points, and $\boldsymbol{\beta}_b$ is a linear combination of the basis functions of the Dirichlet node points. The supports of the basis functions of the Dirichlet node points do not intersect any of the annuli $\Omega_1, \Omega_2, \ldots$, as shown in Figure 7-3a. Therefore, over all of these annuli,

$$\boldsymbol{\beta}_b = 0$$

and

$$\boldsymbol{\beta}_F = \boldsymbol{\beta}_h$$

In a manner similar to the derivation of Equation (6.2.1-14), we have, for Ω_1,

$$\int_{\Omega_1} (\nabla \boldsymbol{\alpha}_i \cdot \nabla \boldsymbol{\beta}_h)\, dv = \int_{S_0} \frac{\partial \gamma}{\partial n_1} \boldsymbol{\alpha}_i\, dS + \int_{S_1} \frac{\partial \gamma}{\partial n_1} \boldsymbol{\alpha}_i\, dS, \qquad 1 \leqslant i \leqslant N_i \qquad (7.3.5\text{-}1)$$

In this equation, the basis functions $1 \leqslant i \leqslant N_i$ are those that are nonzero over Ω_1, and

$$\boldsymbol{\beta}_h = \sum_{i=1}^{N_1} \beta_i \boldsymbol{\alpha}_i \qquad (7.3.5\text{-}2)$$

Let those basis functions corresponding to the node points (a) on S_0, (b) outside S_0 but inside S_1, and (c) on S_1 be elements of the vectors $\hat{\boldsymbol{\alpha}}_0$, $\hat{\boldsymbol{\alpha}}_M$, and $\hat{\boldsymbol{\alpha}}_1$, respectively. (The sum of the elements of these three vectors is then N_1.) Similarly, let $\hat{\beta}_0$, $\hat{\beta}_M$, and $\hat{\beta}_1$ be the vectors that carry the coefficients of $\hat{\boldsymbol{\alpha}}_0$, $\hat{\boldsymbol{\alpha}}_M$, and $\hat{\boldsymbol{\alpha}}_1$ [as in Equation (7.3.5-2)]. Then, from this equation, we have

$$\begin{aligned}\boldsymbol{\beta}_h &= \hat{\boldsymbol{\alpha}}_0^T \hat{\beta}_0 + \hat{\boldsymbol{\alpha}}_M^T \hat{\beta}_M + \hat{\boldsymbol{\alpha}}_1^T \hat{\beta}_1 \\ &= \hat{\beta}_0^T \hat{\boldsymbol{\alpha}}_0 + \hat{\beta}_M^T \hat{\boldsymbol{\alpha}}_M + \hat{\beta}_1^T \hat{\boldsymbol{\alpha}}_1 \end{aligned} \qquad (7.3.5\text{-}3)$$

Substituting the vectors $\hat{\boldsymbol{\alpha}}_0$, $\hat{\boldsymbol{\alpha}}_M$, and $\hat{\boldsymbol{\alpha}}_1$ into Equation (7.3.5-1), we get

$$\int_{\Omega_1} \nabla \hat{\boldsymbol{\alpha}}_0 \cdot \nabla \boldsymbol{\beta}_h\, dv = \int_{S_0} \frac{\partial \gamma}{\partial n_1} \hat{\boldsymbol{\alpha}}_0\, dS \qquad (7.3.5\text{-}4)$$

$$\int_{\Omega_1} \nabla \hat{\boldsymbol{\alpha}}_M \cdot \nabla \boldsymbol{\beta}_h\, dv = 0 \qquad (7.3.5\text{-}5)$$

and

$$\int_{\Omega_1} \nabla \hat{\boldsymbol{\alpha}}_i \cdot \nabla \boldsymbol{\beta}_h \, dv = \int_{S_1} \frac{\partial \gamma}{\partial n_1} \hat{\boldsymbol{\alpha}}_1 \, dS \tag{7.3.5-6}$$

In arriving at these three equations, we used the facts that $\hat{\boldsymbol{\alpha}}_M$ and $\hat{\boldsymbol{\alpha}}_1$ are zero over S_0, and $\hat{\boldsymbol{\alpha}}_0$ and $\hat{\boldsymbol{\alpha}}_M$ are zero over S_1. By substituting Equation (7.3.5-3) into Equations (7.3.5-4), (7.3.5-5), and (7.3.5-6), we obtain

$$A_{00}\hat{\beta}_0 + A_{0M}\hat{\beta}_M + A_{01}\hat{\beta}_1 = \int_{S_0} \frac{\partial \gamma}{\partial n_1} \hat{\boldsymbol{\alpha}}_0 \, dS \tag{7.3.5-7}$$

$$A_{M0}\hat{\beta}_0 + A_{MM}\hat{\beta}_M + A_{M1}\hat{\beta}_1 = 0 \tag{7.3.5-8}$$

$$A_{10}\hat{\beta}_0 + A_{1M}\hat{\beta}_M + A_{11}\hat{\beta}_1 = \int_{S_1} \frac{\partial \gamma}{\partial n_1} \hat{\boldsymbol{\alpha}}_1 \, dS \tag{7.3.5-9}$$

The matrices $A_{00}, A_{0M}, A_{01}, A_{M0}, A_{MM}, A_{M1}, A_{1M}$, and A_{11} in these equations are defined, for example, by

$$A_{M0} = \int_{\Omega_1} \nabla \hat{\boldsymbol{\alpha}}_M \cdot \nabla \hat{\boldsymbol{\alpha}}_0^T \, dv \tag{7.3.5-10}$$

From Equation (7.3.5-8), we have

$$\hat{\beta}_M = -A_{MM}^{-1} A_{M0} \hat{\beta}_0 - A_{MM}^{-1} A_{M1} \hat{\beta}_1 \tag{7.3.5-11}$$

and, when this equation is substituted into Equations (7.3.5-7) and (7.3.5-9), we have

$$B_{00}^{(1)}\hat{\beta}_0 + B_{01}^{(1)}\hat{\beta}_1 = \int_{S_0} \frac{\partial \gamma}{\partial n_1} \hat{\boldsymbol{\alpha}}_0 \, dS \tag{7.3.5-12}$$

and

$$B_{10}^{(1)}\hat{\beta}_0 + B_{11}^{(1)}\hat{\beta}_1 = \int_{S_1} \frac{\partial \gamma}{\partial n_1} \hat{\boldsymbol{\alpha}}_1 \, dS \tag{7.3.5-13}$$

where the matrices B_{00}, B_{01}, B_{10}, and B_{11} are given by

$$B_{00}^{(1)} = A_{00} - A_{0M} A_{MM}^{-1} A_{M0} \tag{7.3.5-14}$$

$$B_{01}^{(1)} = A_{01} - A_{0M} A_{MM}^{-1} A_{M1} \tag{7.3.5-15}$$

$$B_{10}^{(1)} = A_{10} - A_{1M} A_{MM}^{-1} A_{M0} \tag{7.3.5-16}$$

and

$$B_{11}^{(1)} = A_{11} - A_{1M} A_{MM}^{-1} A_{M1} \tag{7.3.5-17}$$

The above equations give the formulation that applies when higher order shape functions are used.

If, on the other hand, lowest order shape functions are used, we can see that Equations (7.3.5-14) through (7.3.5-17) are replaced by

$$B_{00} = A_{00} \tag{7.3.5-18}$$

$$B_{01} = A_{01} \tag{7.3.5-19}$$

$$B_{10} = A_{10} \tag{7.3.5-20}$$

and

$$B_{11} = A_{11} \tag{7.3.5-21}$$

7.3.6. Elimination of Node Points Outside S_0 from Problem

We proceed to write a pair of equations similar to Equations (7.3.5-12) and (7.3.5-13), but for annulus Ω_2. From the fact that Ω_2 is similar to Ω_1 (as discussed above), we can show that the matrices B_{00}, B_{01}, B_{10}, and B_{11} apply to Ω_2 as well. These equations are then

$$B_{00}^{(1)}\hat{\beta}_1 + B_{01}^{(1)}\hat{\beta}_2 = \int_{S_1} \frac{\partial \gamma}{\partial n_2} \hat{\alpha}_1 \, dS \tag{7.3.6-1}$$

and

$$B_{10}^{(1)}\hat{\beta}_1 + B_{11}^{(1)}\hat{\beta}_2 = \int_{S_2} \frac{\partial \gamma}{\partial n_2} \hat{\alpha}_2 \, dS \tag{7.3.6-2}$$

Over S_1, the normal directions n_1 and n_2 are opposite each other, so that

$$\frac{\partial \gamma}{\partial n_1} + \frac{\partial \gamma}{\partial n_2} = 0 \tag{7.3.6-3}$$

If Equations (7.3.5-13) and (7.3.6-1) are added, using this equation, we have

$$B_{10}^{(1)}\hat{\beta}_0 + (B_{11}^{(1)} + B_{00}^{(1)})\hat{\beta}_1 + B_{01}^{(1)}\hat{\beta}_2 = 0$$

which is manipulated to

$$\hat{\beta}_1 = -(B_{00}^{(1)} + B_{11}^{(1)})^{-1}B_{10}^{(1)}\hat{\beta}_0 - (B_{00}^{(1)} + B_{11}^{(1)})^{-1}B_{01}^{(1)}\hat{\beta}_2 \qquad (7.3.6\text{-}4)$$

When β_1 from this equation is substituted into Equations (7.3.5-12) and (7.3.6-2), we have

$$B_{00}^{(2)}\hat{\beta}_0 + B_{01}^{(2)}\hat{\beta}_2 = \int_{S_0} \frac{\partial \gamma}{\partial n_1} \hat{\alpha}_0 \, dS \qquad (7.3.6\text{-}5)$$

and

$$B_{10}^{(2)}\hat{\beta}_0 + B_{11}^{(2)}\hat{\beta}_2 = \int_{S_2} \frac{\partial \gamma}{\partial n_2} \hat{\alpha}_2 \, dS \qquad (7.3.6\text{-}6)$$

where

$$B_{00}^{(2)} = B_{00}^{(1)} - B_{01}^{(1)}(B_{00}^{(1)} + B_{11}^{(1)})^{-1}B_{10}^{(1)} \qquad (7.3.6\text{-}7)$$

$$B_{01}^{(2)} = -B_{01}^{(1)}(B_{00}^{(1)} + B_{11}^{(1)})^{-1}B_{01}^{(1)} \qquad (7.3.6\text{-}8)$$

$$B_{10}^{(2)} = -B_{10}^{(1)}(B_{00}^{(1)} + B_{11}^{(1)})^{-1}B_{10}^{(1)} \qquad (7.3.6\text{-}9)$$

and

$$B_{11}^{(2)} = B_{11}^{(1)} - B_{10}^{(1)}(B_{00}^{(1)} + B_{11}^{(1)})^{-1}B_{01}^{(1)} \qquad (7.3.6\text{-}10)$$

By the above manipulations that take us from Equations (7.3.5-12), (7.3.5-13), (7.3.6-1), and (7.3.6-2) to Equations (7.3.6-5) through (7.3.6-10), we have formulated an annulus that is the union of Ω_1 and Ω_2. And we have eliminated the vector $\hat{\beta}_1$ from the problem. By similar manipulations, we could formulate the union of Ω_3 and Ω_4, obtaining

$$B_{00}^{(2)}\hat{\beta}_2 + B_{01}^{(2)}\hat{\beta}_4 = \int_{S_2} \frac{\partial \gamma}{\partial n_3} \hat{\alpha}_2 \, dS \qquad (7.3.6\text{-}11)$$

$$B_{00}^{(2)}\hat{\beta}_2 + B_{11}^{(2)}\hat{\beta}_4 = \int_{S_4} \frac{\partial \gamma}{\partial n_4} \hat{\alpha}_4 \, dS \qquad (7.3.6\text{-}12)$$

(In the process, we eliminate $\hat{\beta}_3$.) Starting from Equations (7.3.6-5), (7.3.6-6),

(7.3.6-11), and (7.3.6-12), we can, by a similar process, formulate the union of $\Omega_1, \Omega_2, \Omega_3$, and Ω_4, eliminating $\hat{\beta}_2$ in the process. Continuing this process, we can next formulate the union of eight annuli, then sixteen annuli, and so on. In each iteration of this process, we formulate the union of twice as many annuli as before. At the end of, say, k iterations, we will have formulated the union of M annuli, where

$$M = 2^k \tag{7.3.6-13}$$

From Equations (7.3.3-2) and (7.3.6-13), we see that the distance, D_M, from point P to a node point on S_M along some radial is

$$D_M = RK^{2^k} \tag{7.3.6-14}$$

Clearly, the size of the problem domain of our hypothetical problem can increase very rapidly, indeed, after just a few iterations. Furthermore, as we can see from Chapter 3, the scalar potential for which we solve falls off at infinity as the reciprocal of the radius squared. Thus we can see that the exact field, γ, is bounded at infinity by

$$|\gamma| < \frac{C}{D_M^2} = \frac{C}{(RK^{2^k})^2} \tag{7.3.6-15}$$

where C is some positive constant.

After k iterations, then, we have [as an extension of Equations (7.3.6-5) and (7.3.6-6)] that

$$B_{00}^{(k)}\hat{\beta}_0 + B_{01}^{(k)}\hat{\beta}_k = \int_{S_0} \frac{\partial \gamma}{\partial n_1} \hat{\boldsymbol{\alpha}}_0 \, dS \tag{7.3.6-16}$$

$$B_{10}^{(k)}\hat{\beta}_0 + B_{11}^{(k)}\hat{\beta}_k = \int_{S_M} \frac{\partial \gamma}{\partial n_k} \hat{\boldsymbol{\alpha}}_k \, dS \tag{7.3.6-17}$$

From the recursive nature of each iteration, and from Equations (7.3.6-7) through (7.3.6-10), we see that the matrices in these equations can be computed recursively by

$$B_{00}^{(k+1)} = B_{00}^{(k)} - B_{01}^{(k)}(B_{00}^{(k)} + B_{11}^{(k)})^{-1} B_{10}^{(k)} \tag{7.3.6-18}$$

$$B_{01}^{(k+1)} = -B_{01}^{(k)}(B_{00}^{(k)} + B_{11}^{(k)})^{-1} B_{01}^{(k)} \tag{7.3.6-19}$$

$$B_{10}^{(k+1)} = -B_{10}^{(k)}(B_{00}^{(k)} + B_{11}^{(k)})^{-1} B_{10}^{(k)} \tag{7.3.6-20}$$

and

$$B_{11}^{(k+1)} = B_{11}^{(k)} - B_{10}^{(k)}(B_{00}^{(k)} + B_{11}^{(k)})^{-1} B_{01}^{(k)} \tag{7.3.6-21}$$

Suppose, now, that k is the number of the last iteration that we take. Then, over surface S_M, $|\gamma|$ and $|\partial\gamma/\partial n_k|$ are negligibly small, and, to a good approximation, we can say that

$$\hat{\beta}_k = 0 \tag{7.3.6-22}$$

We disregard Equation (7.3.6-17) other than to infer from it, in passing, that $\|B_{10}^{(k)}\|$ must be very small. From Equations (7.3.6-16) and (7.3.6-22) we have

$$B_{00}^{(k)}\hat{\beta}_0 = \int_{S_0} \frac{\partial\gamma}{\partial n_1} \hat{\boldsymbol{\alpha}}_0 \, dS \tag{7.3.6-23}$$

7.3.7. Development of System of Equations for Actual Algorithm

As shown in Figure 7-3, the region R_I includes all space inside the surface S_0 except region R_e, the region that is excluded from the problem domain. As in Section 7.2, we assume that, in general, a portion of the boundary ∂R_e of R_e is ∂e, over which a Dirichlet (and essential) boundary condition is imposed. The remainder of ∂R_e is ∂N, over which a mixed (and natural) boundary condition is imposed. In the previous section, the basis functions of the node points on S_0 are elements of the vector $\hat{\boldsymbol{\alpha}}_0$, and the field values at these node points are elements of the vector $\hat{\beta}_0$. We define similar vectors related to the node points *inside* S_0. There is the vector $\hat{\boldsymbol{\alpha}}_I$ that has as elements the basis functions of these node points. And there is the vector $\hat{\beta}_I$ that has as elements the field values at these points. Then $\boldsymbol{\beta}_h$ is given, inside R_I, by

$$\boldsymbol{\beta}_h = \hat{\beta}_0^T \hat{\boldsymbol{\alpha}}_0 + \hat{\beta}_I^T \hat{\boldsymbol{\alpha}}_I = \hat{\boldsymbol{\alpha}}_0^T \hat{\beta}_0 + \hat{\boldsymbol{\alpha}}_I^T \hat{\beta}_I \tag{7.3.7-1}$$

Using Equation (6.2.1-14), along with the vectors defined above, we have

$$\int_{R_I} \nabla\hat{\boldsymbol{\alpha}}_I \cdot (\nu \nabla \boldsymbol{\beta}_h) \, dv + \int_{\partial N} \sigma \hat{\boldsymbol{\alpha}}_I \boldsymbol{\beta}_h \, dS = \hat{g}_I \tag{7.3.7-2}$$

and

$$\int_{R_I} \nabla\hat{\boldsymbol{\alpha}}_0 \cdot (\nu \nabla \boldsymbol{\beta}_h) \, dv = \hat{g}_0 + \int_{S_0} \frac{\partial\gamma}{\partial n_0} \hat{\boldsymbol{\alpha}}_0 \, dS \tag{7.3.7-3}$$

where

$$\hat{g}_I = \int_{R_I} [\hat{\boldsymbol{\alpha}}_I f - \nabla\hat{\boldsymbol{\alpha}}_I \cdot \nu\nabla\boldsymbol{\beta}_b]\, dv + \int_{\partial N} \hat{\boldsymbol{\alpha}}_I [h - \sigma\boldsymbol{\beta}_b]\, dS \qquad (7.3.7\text{-}4)$$

and

$$\hat{g}_0 = \int_{R_I} \hat{\boldsymbol{\alpha}}_0 f\, dv \qquad (7.3.7\text{-}5)$$

In Equation (7.3.7-3), the derivative $\partial\gamma/\partial n_0$ is taken normally outward over S_0, so that

$$\int_{S_0} \frac{\partial\gamma}{\partial n_0} \hat{\boldsymbol{\alpha}}_0\, dS + \int_{S_0} \frac{\partial\gamma}{\partial n_1} \hat{\boldsymbol{\alpha}}_0\, dS = 0 \qquad (7.3.7\text{-}6)$$

and from this equation and Equations (7.3.6-23) and (7.3.7-3), we have

$$\int_{R_I} \nabla\hat{\boldsymbol{\alpha}}_0 \cdot (\nu\nabla\boldsymbol{\beta}_h)\, dv + B_{00}^{(k)}\hat{\beta}_0 = \hat{g}_0 \qquad (7.3.7\text{-}7)$$

The above equations can now be used to write a matrix-vector equation to be solved for $\hat{\beta}_0$ and $\hat{\beta}_1$. From Equations (7.3.7-1), (7.3.7-2), and (7.3.7-7), we have

$$\begin{bmatrix} (C_{00} + B_{00}^{(k)}) & C_{0I} \\ (C_{I0} + D_{I0}) & (C_{II} + D_{II}) \end{bmatrix} \begin{bmatrix} \hat{\beta}_0 \\ \hat{\beta}_I \end{bmatrix} = \begin{bmatrix} \hat{g}_0 \\ \hat{g}_I \end{bmatrix} \qquad (7.3.7\text{-}8)$$

where the submatrices C_{00}, C_{0I}, C_{I0}, and C_{II} are defined, for example, by

$$C_{0I} = \int_{R_I} \nabla\hat{\boldsymbol{\alpha}}_0 \cdot \nabla(\nu\hat{\boldsymbol{\alpha}}_I^T)\, dv \qquad (7.3.7\text{-}9)$$

and the submatrices D_{I0} and D_{II} are defined, for example, by

$$D_{I0} = \int_{\partial N} \sigma\hat{\boldsymbol{\alpha}}_I\hat{\boldsymbol{\alpha}}_0^T\, dS \qquad (7.3.7\text{-}10)$$

and g_0 and g_I are given by Equations (7.3.7-5) and (7.3.7-4).

7.3.8. Summary of Algorithm

In recapitulation, the algorithm consists of the following steps:

1. Choose a point P near the centroid of the region R_I, as in Figure 7-3, and choose the mapping factor K (say, about 1.5).
2. Cover the region R_I and the annulus Ω_1 with node points, finite elements, shape functions, and basis functions, as shown in Figure 7-3, and from these the vectors $\hat{\boldsymbol{\alpha}}_I$, $\hat{\boldsymbol{\alpha}}_0$, $\hat{\boldsymbol{\alpha}}_1$, and $\hat{\boldsymbol{\alpha}}_M$ (if higher order basis functions are used).
3. Use Equation (7.3.5-10) to compute the matrices A_{00}, A_{01}, A_{10}, and A_{11} (and the matrices A_{0M}, A_{M0}, A_{M1}, A_{1M}, and A_{MM}, if higher order shape functions are used).
4. Compute matrices $B_{00}^{(1)}$, $B_{01}^{(1)}$, $B_{10}^{(1)}$, and $B_{11}^{(1)}$ by Equations (7.3.5-14) through (7.3.5-17) if higher order shape functions are used, or Equations (7.3.5-18) through (7.3.5-21), if lowest order shape functions are used.
5. Compute the matrix $B_{00}^{(k)}$ for a sufficiently large integer k, by the recursive use of Equations (7.3.6-18) through (7.3.6-21).
6. Compute the matrices C_{00}, C_{0I}, C_{II}, D_{I0}, and D_{II} by the use of Equations (7.3.7-9) and (7.3.7-10).
7. Compute the vectors $\hat{g}_0$ and $\hat{g}_I$ using Equations (7.3.7-5) and (7.3.7-4).
8. Compute $\hat{\beta}_0$ and $\hat{\beta}_I$ by solving the matrix-vector equation (7.3.7-8).

As in the McDonald-Wexler algorithm, this long series of steps is not as alarming as it looks, since many of the steps need only be considered once—when writing the computer program.

7.3.9. Error

The error is discussed in terms of the hypothetical algorithm. Since the hypothetical and actual algorithms have the same solution, they have the same error. Furthermore, as shown above, the hypothetical algorithm is, in fact, identical to the algorithm used to solve the interior problem, as discussed in Chapter 6. Therefore, the error norm presented in Section 6.2.4 can be applied to this problem as well.

The problem domain of the hypothetical algorithm, R_M, comprises all space inside the surface, S_M, except the excluded region, R_e.

Suppose that γ is the exact field and that γ_0 is the exact *interior problem* solution we would obtain throughout R_M if we were to require the field to be zero on S_M.

Let

$$\mathbf{e}_M = |\gamma - \gamma_0| \tag{7.3.9-1}$$

We can show, then, that $\mathbf{e}_M$, throughout R_M, is bounded by the maximum of $\mathbf{e}_M$ on S_M. Since

$$\mathbf{e}_M = |\gamma|$$

on S_M, then $\mathbf{e}_M$ is bounded by $|\gamma|$ on S_M. From Inequality (7.3.6-15), we see that it is practical for us to make $|\gamma|$ negligibly small (as pointed out by Silvester in Ref. 3). Therefore, we take $\mathbf{e}_M$ to be negligible.

Since we take the field to be zero over S_M in the hypothetical algorithm, then that algorithm minimizes the norm $\|\gamma_0 - \boldsymbol{\beta}_F\|$, where $\boldsymbol{\beta}_f$ is our finite element solution. This norm is defined [using Equation (6.2.4-2)], by

$$\|\mathbf{e}_0\| = \sqrt{\int_{R_M} \nabla \mathbf{e}_0 \cdot \nu \nabla \mathbf{e}_0 \, dv + \int_{\partial N} \sigma \mathbf{e}_0^2 \, dS} \tag{7.3.9-2}$$

In Equation (7.3.9-2), ∂N is that portion of the boundary of R_e over which the mixed boundary condition is applied. Finally, since we can take $\mathbf{e}_M$ of Equation (7.3.9-1) to be negligible over R_M, then we can, for practical purposes, say that our solution minimizes $\|\gamma - \boldsymbol{\beta}_F\|$ over R_M.

REFERENCES

1. Steele, C. W., "Convergent Algorithm for Unbounded, Two-Dimensional, Linear Eddy Current Problems," COMPUMAG Conference, Grenoble, France, 1978.
2. Chari, M. V. K., and Silvester, P. P., Editors. *Finite Elements in Electrical and Magnetic Field Problems.* New York: John Wiley & Sons, 1980, pp. 179–186.
3. Silvester, P. P., Lowther, D. A., Carpenter, C. J., and Wyatt, E. A., "Exterior Finite Elements for 2-Dimensional Field Problems with Open Boundaries," Proc. IEEE, December 1977, Vol. 124, No. 12.

8
INTEGRAL EQUATION METHOD

8.1. INTRODUCTION

Each field problem done by the integral equation method has the following steps.

1. Using Green's theorem or the Helmholtz theorem (see Section 2.5), develop a suitable expression for the field to be computed as an integral over certain sources. These sources can be real or equivalent currents or charges, or fields or scalar potentials.
2. Express these sources, in turn, in terms of the field to be computed. This leads to an integral equation either in terms of the field or in terms of the sources.
3. Develop a partition of the volume or surface over which the integral is taken. That is, divide this volume or surface suitably into a number of subdivisions.
4. Using this partition and the integral equation, derive a system of equations to be solved for the sources or the field over the volume or surface of integration. If the medium is linear, this is a linear system of equations. If the medium is nonlinear, this is a nonlinear system of equations. In any event, compute the sources over the volume or surface of integration.
5. Using these sources or fields, the partition, and the integral expression developed in step 1, compute the field at any desired point in space.

These steps are exemplified in the sample problem that is worked out in Section 8.9.

In exterior problems, the integral method has the advantage that the field, expressed as an integral over the sources, falls off appropriately at infinity. It has the disadvantage compared to the finite element method in that, for some problems, it requires two major computational steps, steps 4 and 5 above.*

---

*Step 5 is not needed in some computations. For example, to compute the capacitance of a capacitor, Step 4 computes the source (the charge density over its plates) and that is all that is needed.

On the other hand, in the finite element method, there is just one major computational step, which computes the fields at the desired points in space.

The integral equation method can, in general, be applied to a very wide range of problems. It can be used to compute static, quasi-static, and dynamic fields, either vector fields or scalar potentials, in both interior and exterior problems and for a wide variety of media. To discuss all of these applications is beyond the scope of this book. Applications to dynamic problems, for example, can be found in other sources. Our discussion here, however, should certainly enable the reader to extend the method to applications beyond those discussed in this chapter. In this chapter, the applications are limited to:

1. Static problems.
2. Problems with "piecewise continuous" media. That is, if the problem domain is divided into certain subdomains, then the media are continuous throughout those subdomains.
3. Problems that can be solved by computing a scalar potential.
4. Problems in which the medium is isotropic.

In the remainder of this book, the subdomains of 2 above are called "continuity subdomains."

Section 8.2 discusses problems with media that are linear and uniform in the continuity subdomains (which often occur with nonsaturable media). Section 8.3 discusses problems with media that are nonlinear and nonuniform in the continuity subdomains (which often occur with saturable media). Section 8.4 discusses in generality the algorithms used to obtain approximate numerical solutions to the integral equations presented in Sections 8.2 and 8.3. Sections 8.5, 8.6, 8.7, and 8.8 give detailed information needed to implement these algorithms. Finally, Section 8.9 presents a simple sample problem that exemplifies all of the steps needed to define, formulate, construct an algorithm for, and solve a field problem by the integral equation method.

8.2. LINEAR AND UNIFORM MEDIA IN CONTINUITY SUBDOMAINS

8.2.1. General Formulation

This section discusses the formulation of static field problems with linear, uniform, and isotropic media within and throughout each of its continuity subdomains. This problem is typical of configurations in which the medium is unsaturated. The configuration assumed is shown in Figure 8-1. The

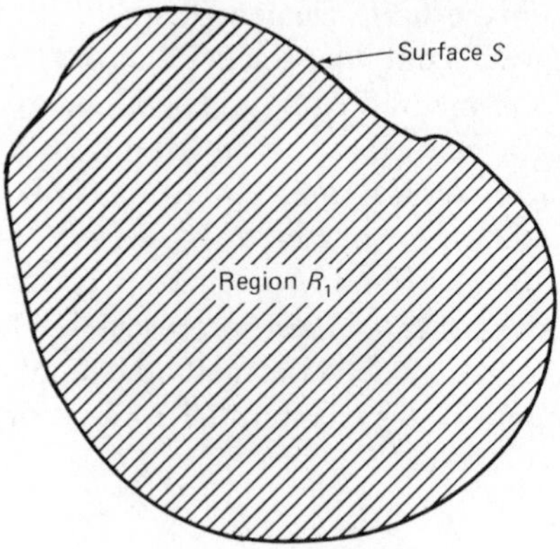

Fig. 8-1. Configuration for field problem with linear and uniform media in continuity subdomains.

problem domain is all of finite space. There are two continuity subdomains, the bounded region R_1 and the unbounded region R_2 (that extends to infinity). They are separated by the surface, *S*. Throughout each of these regions, the medium is uniform, isotropic, and linear.

We assume that there is no electric current or charge throughout finite space. Therefore, either the electric or the magnetic field is conservative, and either of these fields can be represented as the gradient of a total scalar potential. The exact total scalar potentials are γ_1 in R_1 and γ_2 in R_2. All *applied* sources are then at infinity and are represented by an *applied* scalar potential, γ_a, that does not go to zero at infinity. In fact, if R_1 were to go to zero, then we would have

$$\gamma_2 = \gamma_a$$

throughout all of space.

Using equations presented in Chapter 2, it is easy to show that for this configuration,

$$\nabla^2\gamma_1 = \nabla^2\gamma_2 = 0 \tag{8.2.1-1}$$

There can be no real or equivalent sources in either R_1 or finite R_2, but there can be such sources on *S*. The fact that γ_1 and γ_2 both obey Laplace's equation allows us a great deal of flexibility in developing integral equation formulations. As shown in Appendix C, this fact enables us to show that

$$\alpha_1 \gamma_1(\mathbf{r}_f) = K\gamma'_{1S} - L\gamma_{1S} \tag{8.2.1-2}$$

$$(1 - \alpha_1)\gamma_2(\mathbf{r}_f) = -K\gamma'_{2S} + L\gamma_{2S} + \gamma_a \tag{8.2.1-3}$$

where the subscript S indicates quantities taken at the surface S and

$$\alpha_1 = \begin{cases} 1 & \text{if } \mathbf{r}_f \text{ is inside } R_1 \\ \frac{1}{2} & \text{if } \mathbf{r}_f \text{ is on boundary } S \\ 0 & \text{if } \mathbf{r}_f \text{ is inside } R_2 \end{cases} \tag{8.2.1-4}$$

In Equations (8.2.1-2) and (8.2.1-3), K and L are linear operators given by

$$K\gamma'(\mathbf{r}_f) = \int_S \gamma'(\mathbf{r}_s) G(\mathbf{r}_s, \mathbf{r}_f)\, dS \tag{8.2.1-5}$$

$$L\gamma(\mathbf{r}_f) = \int_S \gamma(\mathbf{r}_s)\mathbf{n} \cdot \nabla_s G(\mathbf{r}_s, \mathbf{r}_f)\, dS \tag{8.2.1-6}$$

where, as in Section 2.4,

$$G(\mathbf{r}_s, \mathbf{r}_f) = G_2(\mathbf{r}_s, \mathbf{r}_f) = \frac{1}{2\pi} \ln |\mathbf{r}_s - \mathbf{r}_f|$$

for two-dimensional problems, and

$$G(\mathbf{r}_s, \mathbf{r}_f) = G_3(\mathbf{r}_s, \mathbf{r}_f) = \frac{1}{4\pi |\mathbf{r}_s - \mathbf{r}_f|}$$

for three-dimensional problems, and

$$\gamma'(\mathbf{r}_s) = \frac{\partial}{\partial n} \gamma(\mathbf{r}_s)$$

Since, in the methods of this section, the integrals need to be carried out only over the boundary S, the process is referred to as the "boundary integral method."

We match the Dirichlet and Neumann boundary conditions at the boundary surface, S. Over S, we have

$$\gamma_{1S} = \gamma_{2S} = \gamma_S \tag{8.2.1-7}$$

and

$$\nu_1 \gamma'_{1S} = \nu_2 \gamma'_{2S} \tag{8.2.1-8}$$

If we define an equivalent surface charge density, σ_e, by

$$\sigma_e = \gamma'_{2S} - \gamma'_{1S} \tag{8.2.1-9}$$

and let

$$\nu_r = \frac{\nu_1}{\nu_2} \tag{8.2.1-10}$$

we have

$$\gamma'_{1S} = \frac{\sigma_e}{\nu_r - 1} \tag{8.2.1-11}$$

$$\gamma'_{2S} = \frac{\nu_r \sigma_e}{\nu_r - 1} \tag{8.2.1-12}$$

Then with Equations (8.2.7), (8.2.1-11) and (8.2.1-12) substituted into Equations (8.2.1-2) and (8.2.1-3), we have

$$\alpha_1 \gamma_1 = K\left(\frac{\sigma_e}{\nu_r - 1}\right) - L\gamma_S \tag{8.2.1-13}$$

$$(1 - \alpha_1)\gamma_2 = -K\left(\frac{\nu_r \sigma_e}{\nu_r - 1}\right) + L\gamma_S + \gamma_a \tag{8.2.1-14}$$

When the field point, $\mathbf{r}_f$, is taken on S,

$$\alpha_1 = \tfrac{1}{2}$$

$$\gamma_1 = \gamma_2 = \gamma_S$$

and Equations (8.2.1-13) and (8.2.1-14) become

$$\frac{\gamma_S}{2} = K\left(\frac{\sigma_e}{\nu_r - 1}\right) - L\gamma_S \tag{8.2.1-15}$$

and

$$\frac{\gamma_S}{2} = L\gamma_S - K\left(\frac{\nu_r \sigma_e}{\nu_r - 1}\right) + \gamma_a \tag{8.2.1-16}$$

8.2.2. Exterior Problem

As discussed by Lindholm (Ref. 1), there are a number of ways that we can solve this problem. First, we could solve Equations (8.2.1-15) and (8.2.1-16) simultaneously for the distributions of γ_S and σ_e. Then we could solve for γ_1 (inside S) and γ_2 (outside S) using Equations (8.2.1-13) and (8.2.1-14). Alternatively, we could eliminate σ_e from all four of these equations from the outset. This would give us an integral equation that we could solve for γ_S over S. Then we could solve for γ_1 and γ_2 from the revised versions of Equations (8.2.1-13) and (8.2.1-14). Finally, we could eliminate γ_S from all four equations. This would give us an integral equation that we could solve for σ_e over S. Then we could calculate γ_1 and γ_2 from these revised versions of Equations (8.2.1-13) and (8.2.1-14). Lindholm presents and discusses all of these possibilities in some detail. Just one of these algorithms is presented below. It combines the advantages of being relatively easy to implement and working well even with very large values of ν_r.

Initially the operator K is eliminated from these equations. When Equation (8.2.1-13) is multiplied by ν_r and added to Equation (8.2.1-14), we have

$$\alpha_1 \nu_r \gamma_1 + (1 - \alpha_1)\gamma_2 = \gamma_a - (\nu_r - 1)L\gamma_S \tag{8.2.2-1}$$

When this equation is taken for field points inside S, on S, and outside S, respectively, we have

$$\gamma_1 = \frac{\gamma_a}{\nu_r} - \left(\frac{\nu_r - 1}{\nu_r}\right)L\gamma_S \tag{8.2.2-2}$$

$$\frac{\gamma_S}{2} = \frac{\gamma_a}{\nu_r + 1} - \left(\frac{\nu_r - 1}{\nu_r + 1}\right)L\gamma_S \tag{8.2.2-3}$$

$$\gamma_2 = \gamma_a - (\nu_r - 1)L\gamma_S \tag{8.2.2-4}$$

Notice that as ν_r goes to infinity, Equations (8.2.2-2) and (8.2.2-3) are well behaved, but that the second term on the right of Equation (8.2.2-4) is ill behaved. As ν_r goes to infinity, γ_S becomes more nearly uniform over S. We see from Equation (8.2.2-4) that, when this happens, $L\gamma_S$ tends toward zero.* At the same time, the coefficient of this term, $(\nu_r - 1)$, goes to infinity. These factors can yield unacceptable errors in the computation of γ_2 by Equation (8.2.2-4) if ν_r is very large.

It is necessary to go to an alternative formulation to compute the field outside S. To do this, we add Equations (8.2.1-13) and (8.2.1-14) to obtain

$$\alpha_1\gamma_1 + (1 - \alpha_1)\gamma_2 = -K\sigma_e + \gamma_a \tag{8.2.2-5}$$

and for field points inside S, on S, and outside S, respectively, we have**

$$\gamma_1 = -K\sigma_e + \gamma_a \tag{8.2.2-6}$$

$$\gamma_S = -K\sigma_e + \gamma_a \tag{8.2.2-7}$$

$$\gamma_2 = -K\sigma_e + \gamma_a \tag{8.2.2-8}$$

Notice that all three of these equations are well behaved as ν_r goes to infinity.

The algorithm (that is stable even for very large ν_r) that is proposed by Lindholm is as follows:

1. Solve the integral equation, Equation (8.2.2-3), for γ_S.
2. Use Equation (8.2.2-2) with these values of γ_S to compute γ_1.

---

* If γ_S were uniform over S, with the constant value γ_{SO}, then Equation (8.2.1-6) could be written

$$L\gamma_S = \gamma_{SO}\int_S \mathbf{n}\cdot\nabla_s G(\mathbf{r}_s, \mathbf{r}_f)\,dS$$

By applying the divergence theorem to the vector $\nabla_s G(\mathbf{r}_s, \mathbf{r}_f)$, this equation becomes

$$L\gamma_S = \gamma_{SO}\int_R \nabla_s^2 G(\mathbf{r}_s, \mathbf{r}_f)\,dS$$

where R is the region surrounded by S. Since the field point $\mathbf{r}_f$ is *outside* S for Equation (8.2.2-3), then, throughout R,

$$\nabla_s^2 G(\mathbf{r}_s, \mathbf{r}_f) = 0$$

and, by this and the preceding equation,

$$L\gamma_S = 0$$

** Despite the apparent identity of the right sides of these three equations, they are given as separate equations because, as shown in Equation (8.2.1-5), the operator K depends upon the field point $\mathbf{r}_f$. And this field point is, respectively, inside, on, and outside the surface S in these three equations.

3. Use Equation (8.2.2-7), together with γ_S, as computed in step 1 to solve for σ_e on S.
4. Use Equation (8.2.2-8) with these values of σ_e to compute γ_2.

8.2.3. Interior Problem

In this section we consider the interior field problem in which the medium is linear, uniform, and isotropic. One way to solve this problem is, of course, to use the finite element method. Another way is to use the integral equation method. Suppose, for example, that the problem domain, D, has three dimensions. Then the domain boundary, ∂D, is a surface that has just two dimensions. For the finite element method, our problem domain is the field problem domain, D. But for the integral equation method, our problem domain is the source problem domain, ∂D. Therefore, by the finite element method, we would solve a problem in three dimensions, whereas, by the integral equation method, we would solve a problem in two dimensions. This reduction in dimensionality can have significant advantages. First, it means that we can use the simpler two-dimensional finite elements instead of the more complex three-dimensional finite elements. Second, and more important, we can substantially reduce the number of node points and basis functions used, since these node points need only cover a surface instead of a volume. This is a very important advantage, since the order of the system of equations that we solve equals the number of node points.

We proceed to formulate the interior problem in which certain boundary conditions are imposed on the surface S in Figure 8-1; we then solve for the fields *inside* S. Two such problems are formulated below. First is the problem in which Dirichlet boundary conditions are imposed over all of S (the Dirichlet problem). Second, and more complicated, is the problem in which the Dirichlet boundary condition is imposed over a part of S, and the Neumann boundary condition is imposed over the remainder of S.

For the Dirichlet problem, we first impose continuity of the potential at S by substituting Equation (8.2.1-7) into Equations (8.2.1-2) and (8.2.1-3). Then, assuming no sources at infinity, we let γ_a equal zero. Finally, we then take the difference between these two equations to obtain

$$\alpha_1 \gamma_1 + (1 - \alpha_1)\gamma_2 = K\sigma_e \tag{8.2.3-1}$$

where σ_e is an equivalent surface charge density given by

$$\sigma_e = \gamma'_{1S} - \gamma'_{2S} \tag{8.2.3-2}$$

For the field point $\mathbf{r}_f$ on S, Equation (8.2.3-1) becomes

$$\gamma_S = K\sigma_e \tag{8.2.3-3}$$

And for $\mathbf{r}_f$ inside S, this equation is

$$\gamma_1 = K\sigma_e \tag{8.2.3-4}$$

(Despite appearances, these equations do not imply that γ_S and γ_1 are equal. For the former equation, the Green's function G, used in K [in Equation (8.2.1-5)], is evaluated with $\mathbf{r}_f$ on S. For the latter equation, G is evaluated for $\mathbf{r}_f$ inside S.) Starting with the Dirichlet boundary condition of γ_S defined on S, the algorithm is as follows:

1. Use Equation (8.2.3-3) to calculate an approximation to σ_e on S.
2. Use Equation (8.2.3-4) to calculate γ_1 at all desired points inside S.

We now consider the more complicated problem in which a Dirichlet boundary is imposed over a portion of S (designated ∂E) and a Neumann boundary condition is imposed over the remainder of S (designated ∂N). Our formulation for this problem differs substantially from that of the Dirichlet problem. We start from Equation (8.2.1-2)—

$$\alpha_1\gamma_1 = K\gamma'_{1S} - L\gamma_S$$

—but do not use Equation (8.2.1-3). For the field point $\mathbf{r}_f$ on S, the above equation becomes

$$\frac{\gamma_S}{2} = K\gamma'_{1S} - L\gamma_S \tag{8.2.3-5}$$

and for $\mathbf{r}_f$ inside S, this equation becomes

$$\gamma_1 = K\gamma'_{1S} - L\gamma_S \tag{8.2.3-6}$$

In order to use this equation to compute γ_1, we need to have evaluated *both* γ'_{1S} and γ_S over all of S. At the outset, we know γ_S over ∂E and γ'_{1S} over ∂N. This means, then, that we must compute γ_S over ∂N and γ'_{1S} over ∂E. Section 8.4 presents an algorithm for making these computations numerically.

8.3. SATURABLE, NONLINEAR, AND NONUNIFORM MEDIA IN CONTINUITY SUBDOMAINS

The saturable media problem is characterized by the fact that the constitutive relationships are nonlinear. That is, in an electrostatic problem, D varies nonlinearly with E; and in a magnetostatic problem, B varies nonlinearly with H. Furthermore, the permeability is a function of the magnitude of the magnetic field, and the permittivity is a function of the magnitude of the electric field. These facts make this problem more difficult than the problem with a linear, uniform medium in two important ways.

First, this means, in general, that the permeability or the permittivity varies over the continuity subdomain. For this reason, the equivalent electric charge density given by Equation (2.5.4-12)

$$\rho_e = -\nabla \cdot \mathbf{P}$$

or the equivalent magnetic charge density, given by Equation (2.5.3-9)

$$\rho_m = \nabla \cdot \mathbf{M}$$

(see Section 2.5) is, in general, nonzero over the entire subdomain. (By contrast, in the problems of the previous section these equivalent charge densities were zero over the interiors of the continuity subdomains and nonzero only over their boundaries. Therefore, integrations of sources were needed over only these sub-domain boundaries.) In the problems discussed in this section, we must integrate over the sources that are both on subdomain boundaries and inside them. The need to integrate sources over the interiors of subdomains greatly increases the amount of numerical computation that must be done.

Second, this problem cannot be solved by one simultaneous solution of a system of equations. Take, for example, the magnetostatic problem: the permeability is a function of the magnetic field, but the magnetic field depends, in turn, upon the permeability. This problem then requires an iterative solution in which each iteration produces more accurate distributions of both the permeability and the magnetic field than before.

Trowbridge (Ref. 2, pp. 191–213) discusses such an algorithm.

8.4. NUMERICAL SOLUTION OF INTEGRAL EQUATIONS—GENERAL APPROACH

Section 8.4.1 formulates the general approach to the numerical solution of all linear problems discussed so far in this chapter—with the exception of one problem. That problem is the interior problem with a boundary condition that is part Dirichlet and part Neumann, as discussed in the latter part of Section 8.2.3. The general approach to the numerical solution to that problem is, in turn, discussed in Section 8.4.2.

8.4.1. Typical Problem

We use the approach given in Chapter 5 to formulate numerical solutions to the integral equations given in Section 8.2. The first step is to put the integral equations into operator form. Consider for example, Equation (8.2.2-3) (with terms rearranged).

$$\left(\frac{\nu_r - 1}{\nu_r + 1}\right) L\gamma_S + \frac{\gamma_S}{2} = \frac{\gamma_a}{\nu_r + 1}$$

We define the linear operator, Q, by

$$Q(\mathbf{u}) = \left(\frac{\nu_r - 1}{\nu_r + 1}\right) L\mathbf{u} + \frac{\mathbf{u}}{2} \qquad (8.4.1\text{-}1)$$

and the function $\mathbf{g}$ by

$$\mathbf{g} = \frac{\gamma_a}{\nu_r + 1} \qquad (8.4.1\text{-}2)$$

In these equations, $\mathbf{u}$ and $\mathbf{g}$ are functions defined over the surface S of Figure 8-1. From Equations (8.2.2-3), (8.4.1-1), and (8.4.1-2), we have

$$Q\gamma_S = \mathbf{g} \qquad (8.4.1\text{-}3)$$

which is the operator expression of the integral equation as it was used in Chapter 5 [Equation (5.3-2)] and as it will be used below. As another example, we could put the integral equation, Equation (8.2.2-7),

$$\gamma_S = -K\sigma_e + \gamma_a$$

into operator form by defining Q by

$$Q\mathbf{u} = K\mathbf{u} \qquad (8.4.1\text{-}4)$$

and letting

$$\mathbf{g} = \gamma_a - \gamma_S \qquad (8.4.1\text{-}5)$$

so that Equation (8.2-19) is expressed as

$$Q\sigma_e = \mathbf{g} \qquad (8.4.1\text{-}6)$$

In Chapter 5 the numerical solutions of both differential equations and integral equations were discussed in very general terms. For that discussion, these equations were expressed in the same operator form as Equations (8.4.1-3) and (8.4.1-6). The numerical solution of integral equations is again discussed below in terms of operator equations, but in greater detail. To some extent, this new discussion parallels the discussion given in Chapter 5.

In these discussions, $\boldsymbol{\beta}$, as given by Equation (4.2-1),

$$\boldsymbol{\beta} = \sum_{i=1}^{N} \beta_i \boldsymbol{\alpha}_i$$

is taken to be the numerical approximate solution to the integral equation. That is, if, for example, we were solving Equation (8.4.1-3), then $\boldsymbol{\beta}$ would be an approximation to γ_S. If, alternatively, we were solving Equation (8.4.1-6), then $\boldsymbol{\beta}$ would be an approximation to σ_e. In Equation (4.2-1), $\boldsymbol{\alpha}_i$ is a basis function and β_i is a real number.

We require $\boldsymbol{\beta}$ to satisfy Equation (8.4.1-3) [or Equation (8.4.1-6)] approximately. That is,

$$Q\boldsymbol{\beta} - \mathbf{g} = R \tag{8.4.1-7}$$

where R is the *residual*. By combining this equation with Equation (4.2-1), we have

$$\sum_{i=1}^{N} \beta_i Q \boldsymbol{\alpha}_i - \mathbf{g} = R \tag{8.4.1-8}$$

In the methods of numerical equation solution given below, we produce N Equations (8.4.1-8) by placing some suitable constraint upon R. This gives us an Nth order system of linear equations to solve for $\beta_1, \beta_2, \ldots, \beta_N$.

8.4.2. Formulation for Interior Problem with Dirichlet and Neumann Boundary Conditions

We return to complete the formulation discussed in Subsection 8.2.3 for the solution of the interior problem by the integral equation method, when part of the boundary, ∂E, has a Dirichlet boundary condition, and the remainder of the boundary, ∂N, has a Neumann boundary condition imposed on it. As was pointed out in that section, we must approximate γ_s over ∂N and γ_s' over ∂E, and this requirement means that this problem requires special treatment.

Over the entire boundary, we approximate γ_s by

$$\boldsymbol{\rho}_s = \sum_{i=1}^{N} \rho_i \boldsymbol{\alpha}_i \tag{8.4.2-1}$$

and γ_s' by

$$\boldsymbol{\rho}_s' = \sum_{i=1}^{N} \rho_i' \boldsymbol{\alpha}_i \tag{8.4.2-2}$$

where the ρ_i and the ρ_i' are real numbers. Then, Equation (8.2.3-5) is approximated by

$$\frac{\boldsymbol{\rho}_s}{2} + L\boldsymbol{\rho}_s - K\boldsymbol{\rho}_s' = R \tag{8.4.2-3}$$

Notice that this equation is analogous to Equation (8.4.1-7), which can be solved for a $\boldsymbol{\beta}$ that approximates an exact solution (γ_s or σ_e). When Equations (8.4.2-1) and (8.4.2-2) are substituted into this equation, we have

$$\sum_{i=1}^{N} \left[\rho_i \left(\frac{\boldsymbol{\alpha}_i}{2} + L\boldsymbol{\alpha}_i \right) - \rho_i' K \boldsymbol{\alpha}_i \right] = R \tag{8.4.2-4}$$

We construct our basis functions, $\boldsymbol{\alpha}_i$, so that $\boldsymbol{\alpha}_i$ for $1 \leqslant i \leqslant M$ have their supports in ∂E, and $\boldsymbol{\alpha}_i$ for $M + 1 \leqslant i \leqslant N$ have their supports in ∂N. Since the $\boldsymbol{\alpha}_i$ are all local-support basis functions, this means that the $\boldsymbol{\alpha}_i$ for $1 \leqslant i \leqslant M$ all equal zero over ∂N, and that the $\boldsymbol{\alpha}_i$ for $M + 1 \leqslant i \leqslant N$ all equal zero over ∂E. From this, we see that

$$\sum_{i=1}^{M} \rho_i \boldsymbol{\alpha}_i$$

approximates γ_s over ∂E. Therefore, we can use the Dirichlet boundary condition imposed on γ_s over ∂E to evaluate ρ_i for $1 \leqslant i \leqslant M$. Similarly,

$$\sum_{i=M+1}^{N} \rho_i' \boldsymbol{\alpha}_i$$

approximates γ_s' over ∂N. And we can use the Neumann boundary condition imposed on γ_s' over ∂N to evaluate ρ_i' for $M + 1 \leqslant i \leqslant N$. Using these facts, it is convenient to rewrite Equation (8.4.2-4) as

$$\sum_{i=M+1}^{N} \rho_i \left(\frac{\boldsymbol{\alpha}_i}{2} + L\boldsymbol{\alpha}_i \right) - \sum_{i=1}^{M} \rho_i' K \boldsymbol{\alpha}_i - \mathbf{g} = R \tag{8.4.2-5}$$

where

$$\mathbf{g} = \sum_{i=M+1} \rho_i' K\boldsymbol{\alpha}_i - \sum_{i=1}^{M} \rho_i \left(\frac{\boldsymbol{\alpha}_i}{2} + L\boldsymbol{\alpha}_i \right) \tag{8.4.2-6}$$

Notice that all of the values of ρ_i and ρ_i' that appear in Equation (8.4.2-6) are known and, therefore, $\mathbf{g}$ is known. We wish to solve for the values of ρ_i and ρ_i' that appear in Equation (8.4.2-5). This solution can be simplified by making certain substitutions that put this equation into a somewhat more convenient form. First, we define an operator Q for this problem as follows: For $1 \leqslant i \leqslant M$, so that $\boldsymbol{\alpha}_i$ has its support in ∂E,

$$Q\boldsymbol{\alpha}_i = K\boldsymbol{\alpha}_i \tag{8.4.2-7}$$

and for $1 + M \leqslant i \leqslant N$, so that $\boldsymbol{\alpha}_i$ is in ∂N,

$$Q\boldsymbol{\alpha}_i = \frac{\boldsymbol{\alpha}_i}{2} + L\boldsymbol{\alpha}_i \tag{8.4.2-8}$$

When these equations are substituted into Equation (8.4.2-5), we have

$$\sum_{i=1}^{M} \rho_i' Q\boldsymbol{\alpha}_i - \sum_{i=1+M}^{N} \rho_i Q\boldsymbol{\alpha}_i - \mathbf{g} = R \tag{8.4.2-9}$$

Second, we define β_i as follows: For $1 \leqslant i \leqslant M$,

$$\beta_i = \rho_i' \tag{8.4.2-10}$$

and for $M + 1 \leqslant i \leqslant N$

$$\beta_i = -\rho_i \tag{8.4.2-11}$$

When these equations are substituted into Equation (8.4.2-9), we have

$$\sum_{i=1}^{N} \beta_i Q\boldsymbol{\alpha}_i - \mathbf{g} = R \tag{8.4.2-12}$$

Notice that this equation is identical to Equation (8.4.1-8). Both of these equations are now in a form that is suitable for solution by the methods discussed in the next three sections.

8.5. FINITE ELEMENTS AND BASIS FUNCTIONS USED IN THE INTEGRAL EQUATION METHOD

There are certain special considerations that relate to the use of node points, finite elements, and basis functions as they are used with the integral equation method.

8.5.1. Finite Elements

The finite elements used with the integral equation method must cover all points in the problem domain at which sources or equivalent sources appear.

If, say, the medium is saturable within a certain continuity subdomain, then that medium will, in general, be nonuniform. As a result, the divergence of either **P** (for an electrostatic problem) or **M** (for a magnetostatic problem) will, in general, be nonzero throughout that subdomain. In other words, a *source* will be nonzero throughout the subdomain. In this case, both the boundary *and* the interior of such a subdomain must be covered by finite elements. In a three-dimensional problem, for example, we would require two-dimensional or surface finite elements over the subdomain boundary. We would also require three-dimensional, finite elements over the interior of the subdomain.

Alternatively, if the medium in the continuity subdomain is uniform, linear, and isotropic, then, as discussed in Section 8.2, there are no sources in the interior of the subdomain. The only sources then lie on the boundary of the subdomain. In this case, then, in a three-dimensional problem, we only have two-dimensional finite elements over the boundary S of the subdomain, as shown in Figure 8-1.

The simplest structure of finite elements over this boundary surface would consist of planar triangles. This is quite satisfactory over any portion of the boundary that is in fact a plane surface. However, some boundaries are, at least in part, curved. There are two ways to fit finite elements to curved surfaces. First, we can still use planar triangles to approximate the curved surface. If the sides of the triangle are small compared to the radius of curvature of the boundary, this can be a satisfactory approximation. An alternative is to use *curved* triangles as finite elements. Jeng and Wexler (Ref. 3) show that this can be done with the use of isoparametric finite elements.

8.5.2. Basis Functions

Both pulse and pyramid basis functions are used with the integral equation method to approximate the source distribution. See Chapter 4. Since a source distribution varies continuously (and usually smoothly) over the source problem domain, pyramid basis functions have the advantage that they can represent that source distribution more accurately than pulse basis functions can. On the other hand, pulse basis functions have the advantage of relative simplicity.

8.6. INTEGRAL EQUATION NUMERICAL SOLUTION BY THE COLLOCATION METHOD

8.6.1. Formulation

The first step in the solution of the integral equation by the collocation method is to choose N points p_j, $1 \leqslant j \leqslant N$, distributed throughout the source problem domain (that portion of the problem domain in which sources can exist). For the problem of Section 8.2, the source problem domain is the surface S. For that problem, the points p would be distributed over S. Then we make the residual, R, in Equations (8.4.1-8) and (8.4.2-12) equal to zero at the points p_j so that

$$Q\beta(p_j) = \mathbf{g}(p_j) \qquad (8.6.1\text{-}1)$$

and

$$\sum_{i=1}^{N} \beta_i Q\boldsymbol{\alpha}_i(p_j) = \mathbf{g}(p_j), \qquad 1 \leqslant j \leqslant N \qquad (8.6.1\text{-}2)$$

In Equation (8.6.1-2) the node points of the $\boldsymbol{\alpha}_i$ can be, but do not have to be, coincident with the p_j. It is convenient to put the latter equation in matrix-vector form. We define a matrix M that has the elements

$$m_{ij} = Q\boldsymbol{\alpha}_i(p_j), \qquad 1 \leqslant i, j \leqslant N \qquad (8.6.1\text{-}3)$$

the vector $\hat{\beta}$ that has the elements $\beta_1, \beta_2, \ldots, \beta_N$, and the vector G that has elements $g(p_1), g(p_2), \ldots, g(p_N)$. Then Equation (8.6.1-2) is, in matrix-vector form,

$$M\hat{\beta} = G \qquad (8.6.1\text{-}4)$$

To use the collocation method, we calculate the vector G and the matrix M. Then, assuming that M is nonsingular (as it will be in a well-posed problem), Equation (8.6.1-4) can be solved for $\hat{\beta}$.

8.6.2. Evaluation of Matrix M

The most difficult aspect of an integral equation solution by the collocation method is the computation of the matrix M, with elements given by Equation

(8.6.1-3). This computation is made difficult by the facts that the operator Q involves complicated integrations and that the basis function $\boldsymbol{\alpha}_i$ (that Q operates upon) can, in some cases, be complicated.

In any event, regardless of complexity, these integrations can be accomplished by approximate numerical methods, as discussed later in this chapter. In certain cases, however, these integrations can be carried out analytically. Whether the integrations are done numerically or analytically, we utilize the fact that the integrand of each of these integrals is linear with the basis function. This means that each of these integrals is the sum of integrals taken one at a time over each finite element of the support of the basis function. That is, the complexity of these integrations is reduced to finding the integral over a single finite element.

Analytical algorithms are discussed below for the K and L operators operating on a pulse basis function defined over a triangular finite element, as shown in Figure 4-1a. Lindholm (Ref. 4) and Rao et al. (Ref. 5) present the formulation for K operating on this basis function. As an alternative to Equation (8.2.1-6), we can express the operator L by

$$L\gamma = \frac{1}{4\pi} \int \gamma \, d\Omega \tag{8.6.2-1}$$

where $d\Omega$ is an increment of solid angle subtended by an increment of S, dS, at the field point $\mathbf{r}_f$. We define the function L_T by

$$L_T = \int_{A_T} \beta_p \, d\Omega$$

where β_p is the shape function of the pulse basis function defined over the area of the triangle A_T. Since β_p is a pulse basis function, then

$$\beta_p = 1$$

and

$$L_T = \Omega_T$$

where Ω_T is the solid angle subtended by the triangle A_T. For each side of the triangle, we can define a plane, P_i, that contains that side and passes through the field point $\mathbf{r}_f$. We have then defined the planes P_1, P_2, and P_3 corresponding to the sides of the triangle. We know then (Ref. 6) that

$$L_T = \Omega_T = \theta_1 + \theta_2 + \theta_3 - \pi$$

where θ_1 is the angle between P_1 and P_2, etc. Finally, to compute these angles we have, for example, that

$$\theta_1 = \cos^{-1}(-\mathbf{n}_1 \cdot \mathbf{n}_2)$$

where $\mathbf{n}_1$ and $\mathbf{n}_2$ are unit vectors normal to P_1 and P_2, respectively.

Integrations of higher order shape functions (that is, shape functions with linear and higher order) can be accomplished by numerical methods, as discussed in Section 8.8.

8.7. INTEGRAL EQUATION NUMERICAL SOLUTION BY THE GALERKIN METHOD

8.7.1. General Formulation

As shown in Section 5.4.2, this method involves two steps:

1. We define a Hilbert space, $\mathscr{H}$. The elements of $\mathscr{H}$ are functions defined over the source problem domain. (The term *source problem domain* is defined in Chapter 3. In the problem in Section 8.2 and in Figure 8-1 the source problem domain is the surface S.) The inner product of $\mathscr{H}$ is usually defined as

$$\langle \mathbf{u}, \mathbf{w} \rangle = \int_{D_S} \mathbf{u}\mathbf{w}\, dD_S \tag{8.7.1-1}$$

where $\mathbf{u}$ and $\mathbf{w}$ are elements of $\mathscr{H}$ and D_S is the source problem domain. Thus, for the problem of Section 8.3, we would have

$$\langle \mathbf{u}, \mathbf{w} \rangle = \int_{S} \mathbf{u}\mathbf{w}\, dS \tag{8.7.1-2}$$

2. We then require that

$$\langle R, \boldsymbol{\alpha}_j \rangle = 0, \qquad 1 \leqslant j \leqslant N \tag{8.7.1-3}$$

where R is the residual shown in Equation (8.4.1-7), so that

$$\langle Q\boldsymbol{\beta} - \mathbf{g}, \boldsymbol{\alpha}_j \rangle = 0, \qquad 1 \leqslant j \leqslant N \tag{8.7.1-4}$$

Since we know from Equation (4.2-1) that

$$\boldsymbol{\beta} = \sum_{i=1}^{N} \beta_i \boldsymbol{\alpha}_i$$

we have with Equation (8.7.1-4)

$$\sum_{i=1}^{N} \beta_i \langle Q\boldsymbol{\alpha}_i, \boldsymbol{\alpha}_j \rangle = \langle \mathbf{g}, \boldsymbol{\alpha}_j \rangle, \qquad 1 \leqslant j \leqslant N \tag{8.7.1-5}$$

Again, we define the matrix M as having elements

$$m_{ij} = \langle Q\boldsymbol{\alpha}_i, \boldsymbol{\alpha}_j \rangle \tag{8.7.1-6}$$

and the vectors $\hat{\boldsymbol{\beta}}$ and G as having, respectively, the elements $\beta_1, \beta_2, \ldots, \beta_N$ and the elements $\langle \mathbf{g}, \boldsymbol{\alpha}_1 \rangle, \langle \mathbf{g}, \boldsymbol{\alpha}_2 \rangle, \ldots, \langle \mathbf{g}, \boldsymbol{\alpha}_N \rangle$. Then, in matrix-vector form, Equation (8.7.1-5) is given by

$$M\hat{\boldsymbol{\beta}} = G \tag{8.7.1-7}$$

As discussed in Chapter 5, the basis functions $\boldsymbol{\alpha}_1, \boldsymbol{\alpha}_2, \ldots, \boldsymbol{\alpha}_N$ are chosen so that the matrix M [with elements given by Equation (8.7.1-6)] is nonsingular. From that nonsingularity we know that the linear system of Equation (8.7.1-7) has a unique solution, $\hat{\boldsymbol{\beta}}$.

8.7.2. Integral Equation Numerical Solutions that are Orthogonal Projections

We proceed to examine the circumstances in which the numerical solution of an integral equation is an orthogonal projection. We use the approach given in Section 5.8 for this purpose. As shown in Section 5.8.2, an orthogonal projection has the advantage of minimizing a certain error norm, as described below.

Suppose that we wish to calculate a numerical approximation to $\boldsymbol{\gamma}_s$ as given by Equation (8.4.1-3). Combining this equation with Equation (8.7.1-4) gives us

$$\langle Q(\boldsymbol{\gamma}_s - \boldsymbol{\beta}), \boldsymbol{\alpha}_j \rangle = 0, \qquad 1 \leqslant j \leqslant N \tag{8.7.2-1}$$

If Q is a symmetric, positive-definite operator, then, as shown in Section 5.8, we can define a new inner product $\langle \cdot \rangle_0$, so that

$$\langle \mathbf{u}, \mathbf{w} \rangle_0 = \langle Q\mathbf{u}, \mathbf{w} \rangle \tag{8.7.2-2}$$

Then, combining Equations (8.7.2-1) and (8.7.2-2) gives us

$$\langle \boldsymbol{\gamma}_s - \boldsymbol{\beta}, \boldsymbol{\alpha}_j \rangle_0 = 0, \qquad 1 \leqslant j \leqslant N \tag{8.7.2-3}$$

Since $\boldsymbol{\beta}$ is a linear combination of $\boldsymbol{\alpha}_1, \boldsymbol{\alpha}_2, \ldots, \boldsymbol{\alpha}_N$ [as shown in Equation (4.2-1)], this equation shows us that $\boldsymbol{\beta}$ is the orthogonal projection of $\boldsymbol{\gamma}_s$ onto an N-dimensional subspace of $\mathscr{H}$ that has as a basis the functions $\boldsymbol{\alpha}_1, \boldsymbol{\alpha}_2, \ldots, \boldsymbol{\alpha}_N$. Therefore, as shown in Section 5.8.2, this orthogonal projection, $\boldsymbol{\beta}$, minimizes a norm of the error, specifically,

$$\|\boldsymbol{\gamma}_s - \boldsymbol{\beta}\|_0 = \sqrt{\langle \boldsymbol{\gamma}_s - \boldsymbol{\beta}, \boldsymbol{\gamma}_s - \boldsymbol{\beta} \rangle_0} \tag{8.7.2-4}$$

Furthermore, since Q is symmetric and positive-definite, and since $\boldsymbol{\beta}$ is an orthogonal projection, this method is an orthogonal projection method, as discussed in Section 5.8.4.

Summing up, if the operator Q is symmetric and positive-definite, then:

1. We can define a new inner product, $\langle \cdot \rangle_0$, as given by Equation (8.7.2-2).
2. This method minimizes the natural norm of this inner product of the error $\|\gamma_s - \beta\|_0$ as given by Equation (8.7.2-4);
3. This method is known as an orthogonal projection method.

Similar statements could be made if $\boldsymbol{\beta}$ had, instead, been an approximation to the equivalent charge density, σ_e, as given by Equation (8.4.1-6).

The truth of statements 1, 2, and 3 above depends simply upon whether or not the operator Q is symmetric and positive-definite. This may, in some cases, be a difficult question. We can, however, answer this question for Q as defined by Equations (8.4.1-1) and (8.4.1-4)

First, using Equations (8.2.1-6) and (8.7.1-1), we see that

$$\langle L\mathbf{u}, \mathbf{w} \rangle = \int_S \mathbf{w}(\mathbf{r}_f) \int_S \mathbf{u}(\mathbf{r}_s) \frac{\partial}{\partial n_s} G(\mathbf{r}_s, \mathbf{r}_f)\, dS_s\, dS_f \tag{8.7.2-5}$$

whereas

$$\langle L\mathbf{w}, \mathbf{u} \rangle = \int_S \mathbf{u}(\mathbf{r}_f) \int_S \mathbf{w}(\mathbf{r}_s) \frac{\partial}{\partial n_s} G(\mathbf{r}_s, \mathbf{r}_f)\, dS_s\, dS_f \tag{8.7.2-6}$$

In these equations, we see that the term

$$\frac{\partial}{\partial n_s} G(\mathbf{r}_s, \mathbf{r}_f)$$

involves taking the derivative of the Green's function normal to S at the source point $\mathbf{r}_s$. In order for the integrals of Equations (8.7.2-5) and (8.7.2-6) to be equal, this normal derivative would have to be, in general, equal to

$$\frac{\partial}{\partial n_f} G(\mathbf{r}_s, \mathbf{r}_f)$$

where this derivative is taken normal to S at the *field* point, $\mathbf{r}_f$. In general, these two normal derivatives are not equal. Therefore, in general,

$$\langle L\mathbf{u}, \mathbf{w}\rangle \neq \langle L\mathbf{w}, \mathbf{u}\rangle \tag{8.7.2-7}$$

that is, the operator L is, in general, nonsymmetric. Then the operator Q, as defined in Equation (8.4.1-1),

$$Q\mathbf{u} = \left(\frac{\nu_r - 1}{\nu_r + 1}\right) L\mathbf{u} + \frac{\mathbf{u}}{2}$$

is, in general, nonsymmetric. Therefore, a numerical solution calculated using this operator will, in general, not be an orthogonal projection.

On the other hand, Wexler (Ref. 7) has shown that the operator K, given by Equation (8.2.1-5), is both symmetric and positive-definite. That is, using the operator K in Equation (8.2.1-5) and the inner product of Equation (8.7.1-1), we have

$$\langle K\mathbf{u}, \mathbf{w}\rangle = \langle K\mathbf{w}, \mathbf{u}\rangle \tag{8.7.2-8}$$

and

$$\langle K\mathbf{u}, \mathbf{u}\rangle > 0 \tag{8.7.2-9}$$

if

$$\mathbf{u} \neq 0$$

From this we can easily show that the operator Q, as given by Equation (8.4.1-4),

$$Q\mathbf{u} = K\mathbf{u}$$

is symmetric and positive-definite. Therefore, a numerical solution, β, calculated using this Q will be an orthogonal projection.

8.7.3. Evaluation of Matrix *M*

As with the collocation method, the most difficult aspect of an integral equation solution by the Galerkin method is the computation of the matrix M, with elements given by Equation (8.7.1-6). From Equations (8.7.1-1) and (8.7.1-6) we see that

$$m_{ij} = \int_{D_S} (Q\alpha_i)\alpha_j \, dD_S \qquad (8.7.3\text{-}1)$$

Since Q is an integral operator, we see that the computation of m_{ij} will require a double integration. These integrations are accomplished by numerical integration as discussed in Section 8.8 and Reference 3.

8.8. NUMERICAL INTEGRATION

8.8.1. General Considerations

As shown in the previous two sections, the integrations required for implementation of the integral equation method are sometimes simple enough that they can be performed analytically. Alternatively, these integrations are frequently so difficult that analytical integration is either very difficult or impossible. In these cases, numerical integration is used instead.

The most popular method of numerical integration as applied to numerical solutions of integral equations is the Gauss-Quadrature method. For an understanding of the basic nature of the Gauss-Quadrature method, the reader is referred to a discussion of its application to integration over a one-dimensional domain (Ref. 8, pp. 330–332). We see that in that case, for a given degree of accuracy, one can sum the integrand over fewer points than one would use in the Newton-Cotes method of numerical integration (Ref. 8, pp. 328–329). The only price that one pays for the use of fewer points is that, in the Newton-Cotes method, the points (at which the integrand is evaluated) are spaced equally over the domain of integration; alternatively, in the Guass-Quadrature method, these points are taken at certain prescribed points within the domain. It is commonly thought that the somewhat more complicated algorithm for the placement of these points is a small price to pay for the ability to use fewer points for a given accuracy.

Again, with integration over a two-dimensional domain or a three-dimensional domain, we can have numerical integration by either the Newton-Cotes method or by the Gauss-Quadrature method. And, again, for two-dimensional or three-dimensional integration, Gauss-Quadrature integration is preferred for the reasons given above. In the remainder of this section, single integrations over rectangles and triangles by the Gauss-Quadrature method are discussed. (Similar integration techniques apply to three-dimensional finite elements such as tetrahedra and rectangular boxes.) We will see that these Gauss-Quadrature formulas for the integral over any rectangle and triangle (or, for that matter, the Gauss-Quadrature formulas for integration over any other finite element) can be expressed in the form

$$I_a = A \sum_{i=1}^{N} W_i f(P_i) \tag{8.8.1-1}$$

In this equation, I_a is the Gauss-Quadrature of the integral over a finite element, P_i is a Gauss-Quadrature interpolation point, and N is the total number of those points over the finite element. The area or volume of the finite element, depending upon whether it is two-dimensional or three-dimensional, is A, while, W_i is a weighing function and $f(P_i)$ is the integrand evaluated at the point P_i. Finally, numerical methods are extended to double integrations, as required in the method of weighted residuals (or the Galerkin method).

8.8.2. Gauss-Quadrature Integration Over a Rectangle

Suppose that we have the double integral

$$I = \int_{y_0}^{y_u} \int_{x_0}^{x_u} f(x, y)\, dx\, dy \tag{8.8.2-1}$$

That is, the function $f(x, y)$ is to be integrated over a rectangle for which

$$x_0 \leqslant x \leqslant x_u$$

$$y_0 \leqslant y \leqslant y_u$$

We proceed to express the numerical integration in terms of the Gauss-Quadrature formula for integration over a one-dimensional domain. We let

$$g(y) = \int_{x=x_0}^{x_u} f(x, y)\, dx \tag{8.8.2-2}$$

so that, with Equation (8.8.2-1),

$$I = \int_{y=y_0}^{y_u} g(y)\, dy \tag{8.8.2-3}$$

Using these equations, we can derive the following equations for expressing I_a, the Gauss-Quadrature approximation to I. First, from Equation (8.8.2-3),

$$I_a = (y_u - y_0) \sum_{i=1}^{n} v_i g_a(y_i) \tag{8.8.2-4}$$

and from Equation (8.8.2-2),

$$g_a(y_i) = (x_u - x_0) \sum_{j=1}^{m} u_j f(x_j, y_i) \tag{8.8.2-5}$$

In these equations $g_a(y_i)$ is a Gauss-Quadrature approximation to $g(y_i)$, and u_j and v_i are Guass-Quadrature weighting functions (Ref. 7, pp. 328–329). Furthermore, the x_j for $1 \leqslant j \leqslant m$ are the corresponding Gauss-Quadrature interpolating points over the interval $x_0 \leqslant x \leqslant x_u$. And the y_i for $1 \leqslant i \leqslant n$ are the corresponding Gauss-Quadrature interpolating points over the interval $y_0 \leqslant y \leqslant y_u$.

8.8.3. Gauss-Quadrature Integration over a Triangle

Both asymmetrically located and symmetrically located Gauss-Quadrature points over a triangle have been computed and published. Zienkiewicz has published asymmetrically located points (Ref. 9), and Vichnevelsky has published symmetrically located points (Ref. 8, p. 335). Of these, the symmetrically located points seem to be more popular, and these are chosen for discussion below. In the symmetrical configurations, the approximate or Gauss-Quadrature integral, I_a, is expressed as

$$I_a = A \sum_{i=1}^{m} v_i \sum_{j=1}^{n_i} f(P_{ij}) \tag{8.8.2-6}$$

In this equation, A is the area of the triangle, v_i is a weighting function, and m is simply the number of such weighting functions used in this formula. The P_{ij} are the Gauss-Quadrature interpolating points, and n_i is the number of interpolating points used for a given weighting function, w_i. The P_{ij} are expressed as functions of the triangular area coordinates discussed in Chapter 4. For a given index i, these P_i are symmetrically located *in terms of these area coordinates.* Reference 7 gives a tabulation of the v_i and the P_{ij} for a large number of symmetric Gauss-Quadrature formulas.

For the simplest (and most heavily used) of these formulas, $m = 1$ and $n_1 = 3$, so that Equation (8.8.2-6) becomes

$$I_a = Av_i \sum_{j=1}^{3} f(P_{ij}) \tag{8.8.2-7}$$

The interpolation points of the form

$$P_{ij} = (\Lambda_1^{(j)}, \Lambda_2^{(j)}, \Lambda_3^{(j)})$$

are given by

$$P_{11} = (0.666666, 0.166666, 0.16666)$$

$$P_{12} = (0.166666, 0.666666, 0.166666)$$

$$P_{13} = (0.166666, 0.166666, 0.666666)$$

and

$$v_1 = \tfrac{1}{3} \tag{8.8.2-8}$$

8.8.3. Double Numerical Integrations over the Problem Domain

As pointed out in Section 8.7.3, it is necessary, in the solution of an integral equation by the Galerkin method to carry out a double integration, each integration being over part or all of the problem domain. Furthermore, for certain integrands that we encounter in practice, it is practically necessary for both of these to be numerical integrations. In these cases, we typically carry out both of these numerical integrations by the Gauss-Quadrature method, as discussed below. Specifically, we may choose this approach to evaluate each element m_{ij} of the matrix M, as given by Equation (8.7.3-1):

$$m_{ij} = \int_{D_s} Q\alpha_i\alpha_j \, dS$$

To understand this integral, we recognize certain of its properties. First, since α_j is a local-support basis function, it equals zero at all points in D_s outside the support of α_j [given by $D_s(j)$]. Second, it is the field point, $\mathbf{r}_f$ that is varied over $D_s(j)$ to accomplish the integration. This equation then becomes

$$m_{ij} = \int_{D_s(j)} Q\alpha_i(\mathbf{r}_f)\alpha_j(\mathbf{r}_f) \, dS \tag{8.8.3-1}$$

Suppose that there are N Gauss-Quadrature points over the finite elements in $D_s(j)$ and that the field point values at these points are $\mathbf{r}_f^{(1)}, \mathbf{r}_f^{(2)}, \ldots, \mathbf{r}_f^{N}$. Then, using Equation (8.8.1-1) to give the Gauss-Quadrature approximation to this integral, we have

$$m_{ij} = A(j) \sum_{k=1}^{N} w_k Q\boldsymbol{\alpha}_i(\mathbf{r}_f^{(k)})\boldsymbol{\alpha}_j(\mathbf{r}_f^{(k)}) \qquad (8.8.3\text{-}2)$$

where $A(j)$ is the area or volume of the domain $D_s(j)$. In this equation, $\boldsymbol{\alpha}_j(\mathbf{r}_f^{(k)})$ is simply the basis function $\boldsymbol{\alpha}_j$ evaluated at the field (and Gauss-Quadrature) point $\mathbf{r}_f^{(k)}$ and is easily evaluated. However, the operator $Q\boldsymbol{\alpha}_i(\mathbf{r}_f^{(k)})$ involves another integration. In this integration, the Gauss-Quadrature points will be *source points* as shown, for example, in the definitions of the L and K operators in Equations (8.2.1-5) and (8.2.1-6).

8.9. SAMPLE PROBLEM

A rather simple sample problem is worked out in this section in order to give the reader a feeling for how the information presented in Chapters 2, 3, 4, and 5, as well as Sections 8.2 and 8.3, fit together to form the basis for working out a problem by the integral equation method.

8.9.1. Problem Definition

A two-dimensional exterior problem is chosen since, as we have seen, the integral equation method is well suited to exterior problems. Figure 8-2 shows the configuration. All sources are at infinity, causing an applied potential field γ_a, that is proportional to the y coordinate in this figure. In this configuration, there is a permeable cylinder, with its axis perpendicular to the paper. The permeability in this cylinder is uniform, linear and isotropic. The problem is to compute the static total scalar magnetic potential both inside and outside the cylinder, that is, throughout all of space.

Again, we divide all of space into two regions. Region R_1 comprises all space inside the cylindrical surface, S_i; and Region R_2 comprises all space outside the cylinder. As in Section 8.2, there are no sources (equivalent or otherwise) in either R_1 or R_2. Therefore, the scalar potential obeys Laplace's equation in both of these regions, and the algorithms of Section 8.2 can be used. Furthermore, all equivalent sources lie on the surface, S.

An advantage of this problem is that it can easily be solved analytically. The potentials γ_1 and γ_2 in R_1 and R_2 respectively, are given by

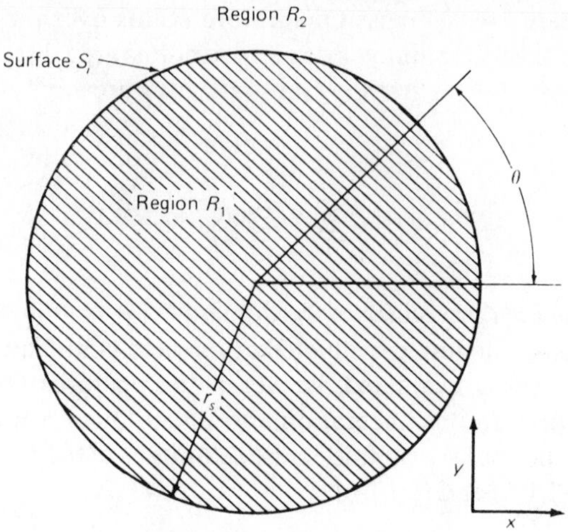

Fig. 8-2. Configuration for sample problem.

$$\gamma_1 = \frac{2\gamma_a r \sin\theta}{\mu_r + 1} \tag{8.9.1-1}$$

and

$$\gamma_2 = \gamma_a r \left[1 - \left(\frac{\mu_r - 1}{\mu_r + 1} \right) \left(\frac{r_s}{r} \right)^2 \right] \sin\theta \tag{8.9.1-2}$$

where,

$$\mu_r = \frac{\mu_1}{\mu_2}$$

The polar coordinates are represented by r and θ, and r_s is the radius of the cylinder, as shown in Figure 8-2. As shown below, this solution enables us to gain some measure of the accuracy of our approximate numerical solution by the integral equation method.

8.9.2. Choice of Basic Formulation

Since potentials in both R_1 and R_2 satisfy Laplace's equation, this problem lends itself to the formulation discussed in Section 8.2. Among the formulations presented there are Equations (8.2.2-2), (8.2.2-3), and (8.2.2-4). These equations have the following two advantages:

1. They involve γ_s but *not* σ_e on the surface S.
2. They involve the operator L but *not* the operator K.

As discussed in that section, the only problem here is with Equation (8.2.2-4) (for γ_2), in the case in which μ_r is very large. To avoid this difficulty, in this sample problem, we will choose a value of μ_r that is not large.

8.9.3. Algorithm

As with the problems discussed in Section 8.2, the problem domain for computing γ_S is the surface, S, in Figure 8-3. This surface is divided into N finite elements, which are identical segments of the cylinder, as shown in this figure. For this problem, the basis functions $\boldsymbol{\alpha}_i$ for $1 \leqslant i \leqslant N$ are pulse basis functions. We obtain our approximate solution on S, $\boldsymbol{\beta}_S$, which approximates γ_S, by the collocation method, with collocation points $\mathbf{p}_j$, $1 \leqslant j \leqslant N$, as shown in Figure 8-3. Since we use the collocation method, $\boldsymbol{\beta}_S$ satisfies Equation (8.2.2-3) exactly at these points $\mathbf{p}_j$. That is,

$$\frac{1}{2}\boldsymbol{\beta}_S(\mathbf{p}_j) = \frac{\gamma_a(\mathbf{p}_j)}{\mu_r + 1} - \left(\frac{\mu_r - 1}{\mu_r + 1}\right) L\boldsymbol{\beta}_S(\mathbf{p}_j) \tag{8.9.3-1}$$

Again, $\boldsymbol{\beta}_S$ is given by

$$\boldsymbol{\beta}_S = \sum_{i=1}^{N} \beta_i \boldsymbol{\alpha}_i \tag{8.9.3-2}$$

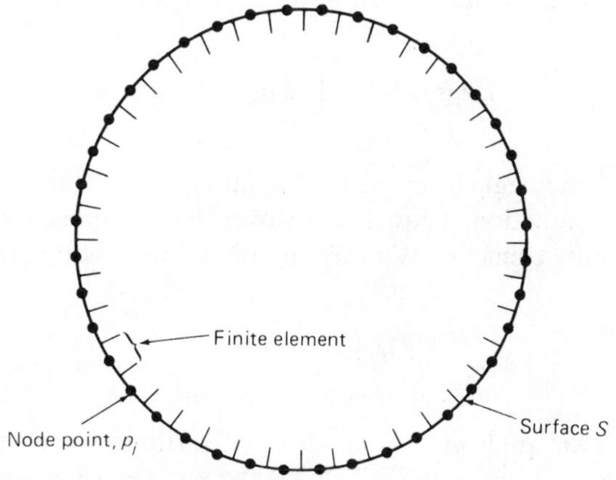

Fig. 8-3. Configuration of node points and finite elements for sample problem.

where the β_i are real numbers. After $\boldsymbol{\beta}_S$ has been computed, we can then solve for $\boldsymbol{\beta}_1$ (the approximation to γ_1 in R_1) and $\boldsymbol{\beta}_2$ (the approximation to γ_2 in R_2). The equations for these fields are found from Equations (8.2.2-2) and (8.2.2-4) to be

$$\boldsymbol{\beta}_1(\mathbf{r}_f) = \frac{\gamma_a(\mathbf{r}_f)}{\mu_r} - \left(\frac{\mu_r - 1}{\mu_r}\right) L\boldsymbol{\beta}_S(\mathbf{r}_f) \tag{8.9.3-3}$$

and

$$\boldsymbol{\beta}_2(\mathbf{r}_f) = \gamma_a(\mathbf{r}_f) - (\mu_r - 1) L\boldsymbol{\beta}_S(\mathbf{r}_f) \tag{8.9.3-4}$$

The next step in the development is to formulate the function $L\boldsymbol{\beta}_S$, since it occurs in the expressions for $\boldsymbol{\beta}_1$, $\boldsymbol{\beta}_2$, and $\boldsymbol{\beta}_S$. Since, from Equation (8.9.3-2), we have

$$L\boldsymbol{\beta}_S(\mathbf{r}_f) = \sum_{i=1}^{N} \beta_i L\boldsymbol{\alpha}_i(\mathbf{r}_f)$$

this means working out an expression for $L\boldsymbol{\alpha}_i(\mathbf{r}_f)$ for any field point, $\mathbf{r}_f$. Since this is a two-dimensional problem, the two-dimensional Green's function is used with Equation (8.2.1-6), and we have

$$L\boldsymbol{\alpha}_i(\mathbf{r}_f) = \frac{1}{2\pi} \int_S \boldsymbol{\alpha}_i(\mathbf{r}_s) \mathbf{n} \cdot \nabla_s \ln |\mathbf{r}_s - \mathbf{r}_f| \, dS$$

From this equation, we can show that

$$L\boldsymbol{\alpha}_i(\mathbf{r}_f) = \frac{1}{2\pi} \int_S \boldsymbol{\alpha}_i(\mathbf{r}_s) \, d\theta(\mathbf{r}_f, \mathbf{r}_s)$$

where $\theta(\mathbf{r}_f, \mathbf{r}_s)$ is the angle from the field point $\mathbf{r}_f$ to the source point $\mathbf{r}_s$. Since $\boldsymbol{\alpha}_i$ is a pulse basis function, it equals unity over the ith finite element, and zero over all other finite elements. With this in mind, this equation becomes

$$L\boldsymbol{\alpha}_i(\mathbf{r}_f) = \frac{1}{2\pi} \Psi(\mathbf{r}_f, i) \tag{8.9.3-5}$$

where $\Psi(\mathbf{r}_f, i)$ is the angle at $\mathbf{r}_f$ subtended by the ith finite element.

By combining Equations (8.9.3-1) through (8.9.3-5), we get our field equations in a form that is much more usable for numerical computation.

$$\frac{1}{2}\boldsymbol{\beta}_S(\mathbf{p}_j) = \frac{\gamma_a(\mathbf{p}_j)}{\mu_r + 1} - \frac{1}{2\pi}\left(\frac{\mu_r - 1}{\mu_r + 1}\right)\sum_{i=1}^{N} \beta_i \Psi(\mathbf{p}_j, i), \qquad 1 \leqslant j \leqslant N \quad (8.9.3\text{-}6)$$

$$\boldsymbol{\beta}_1(\mathbf{r}_f) = \frac{1}{\mu_r}\gamma_a(\mathbf{r}_f) - \frac{1}{2\pi}\left(\frac{\mu_r - 1}{\mu_r}\right)\sum_{i=1}^{N} \beta_i \Psi(\mathbf{r}_f, i) \quad (8.9.3\text{-}7)$$

$$\boldsymbol{\beta}_2(\mathbf{r}_f) = \gamma_a(\mathbf{r}_f) - \frac{1}{2\pi}(\mu_r - 1)\sum_{i=1}^{N} \beta_i \Psi(\mathbf{r}_f, i) \quad (8.9.3\text{-}8)$$

We proceed to formulate the solution for the real number coefficients $\beta_1, \beta_2, \ldots, \beta_N$ by solving the system of Equations (8.9.3-6). This method follows the approach given in Section 8.6.1. Using Equation (8.9.3-2), we have

$$\boldsymbol{\beta}_S(\mathbf{p}_j) = \sum_{i=1}^{N} \beta_i \boldsymbol{\alpha}_i(\mathbf{p}_j), \qquad 1 \leqslant j \leqslant N \quad (8.9.3\text{-}9)$$

We know that the basis function $\boldsymbol{\alpha}_i$ equals unity at the collocation point that falls within its own support and that it equals zero at all other collocation points. That is,

$$\boldsymbol{\alpha}_i(\mathbf{p}_j) = \delta_{ij}, \qquad 1 \leqslant i, j \leqslant N \quad (8.9.3\text{-}10)$$

where δ_{ij} is the Kronacker delta, given by

$$\delta_{ij} = 0, \qquad i \neq j$$

$$\delta_{ii} = 1$$

From this, we see that

$$\boldsymbol{\beta}_S(\mathbf{p}_j) = \beta_j \quad (8.9.3\text{-}11)$$

By combining Equations (8.9.3-6) and (8.9.3-11), we have

$$\sum_{i=1}^{N} \beta_i \left[\frac{1}{2}\delta_{ij} + \frac{1}{2\pi}\left(\frac{\mu_r - 1}{\mu_r + 1}\right)\Psi(\mathbf{p}_j, i)\right] = \frac{1}{\mu_r + 1}\gamma_a(\mathbf{p}_j), \qquad 1 \leqslant j \leqslant N \quad (8.9.3\text{-}12)$$

In this case, the vector G of Equation (8.6.1-4) has the elements

$$g(\mathbf{p}_j) = \frac{1}{\mu_r + 1} \Upsilon_a(\mathbf{p}_j), \qquad 1 \leqslant j \leqslant N \tag{8.9.3-13}$$

The elements m_{ij} of the matrix M of Equation (8.6.1-4) are given from Equation (8.9.3-12) by

$$m_{ij} = \frac{1}{2}\delta_{ij} + \frac{1}{2\pi}\left(\frac{\mu_r - 1}{\mu_r + 1}\right)\Psi(\mathbf{p}_j, i), \qquad 1 \leqslant i, j \leqslant N \tag{8.9.3-14}$$

The algorithm for this problem is then as follows:

1. Compute the vector G and the matrix M by means of Equations (8.9.3-13) and (8.9.3-14).
2. Compute the vector $\hat{\beta}$ (with elements $\beta_1, \beta_2, \ldots, \beta_N$) by solving the Nth order matrix-vector Equation (8.6.1-4).

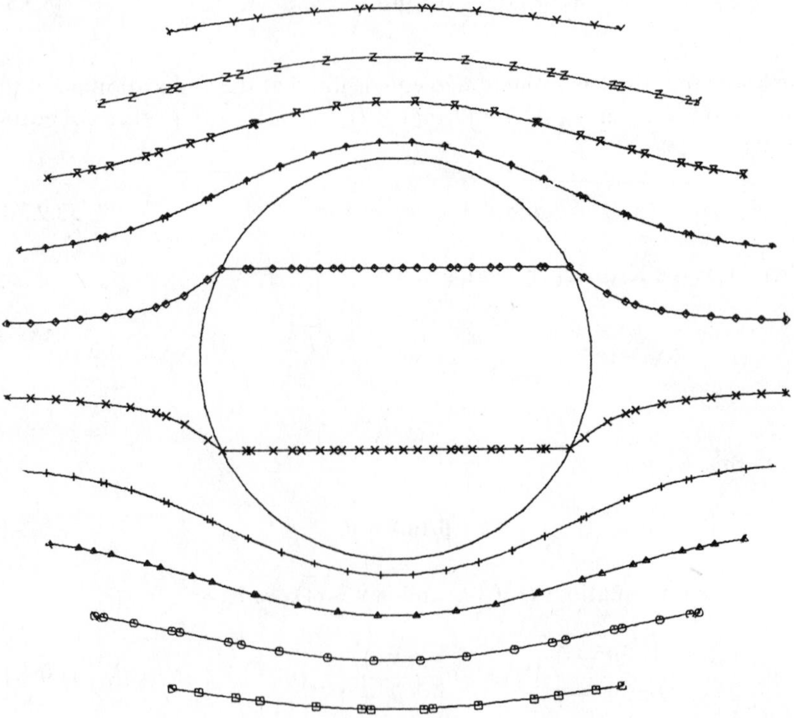

Fig. 8-4. Magnetic potential for sample problem.

3. Compute $\boldsymbol{\beta}_1(\mathbf{r}_f)$ at all desired field points $\mathbf{r}_f$ in R_1 by Equation (8.9.3-7).
4. Compute $\boldsymbol{\beta}_2(\mathbf{r}_f)$ at all desired field points $\mathbf{r}_f$ in R_2 by Equation (8.9.3-8).

8.9.4. Computed Results

Figure 8-4 shows the configuration and contours of constant scalar magnetic potential. These magnetic scalar potential values were computed using the integral equation method algorithm that was presented in the previous section. In this problem, the medium in R_1 [the cylinder, as shown in Figure 8-2] has a permeability of 5. The surface S of the cylinder was divided into 40 finite elements, which appear as identical arcs on Figure 8-3. A field computation was also made using the analytically derived Equations (8.9.1-1) and (8.9.1-2). When these analytically computed fields were contour-plotted, the resulting plot was identical to Figure 8-4.

REFERENCES

1. Lindholm, Dennis, "Notes on Boundary Integral Equations for Three-Dimensional Magnetostatics," *IEEE Transaction on Magnetics*, November 1980, Vol. MAG-16, No. 6, pp. 1409–1413.
2. Chari, M. V. K., and Silvester, P. P., Editors. *Finite Elements in Electrical and Magnetic Field Problems.* New York: John Wiley & Sons, 1980.
3. Jeng, G., and Wexler, A., "Isoparametric, Finite Element Variational Solution of Integral Equations for Three-Dimensional Fields," *International Journal for Numerical Methods in Engineering*, Vol. II, 1977, pp. 1455–1471.
4. Lindholm, Dennis, "Effect of Track Width and Side Shields on the Long Wavelength Response of Rectangular Magnetic Heads," *IEEE Transactions on Magnetics*, March 1980, Vol. MAG-16, No. 2, pp. 430–435.
5. Rao, S. M., Glisson, A. W., Wilton, D. R., and Vidula, B. S., "A Simple Numerical Solution Procedure for Statics Problems Involving Arbitrary- Shaped Surfaces," *IEEE Transactions on Antennas and Propagation*, September 1979, Vol. AP-27, No. 5, pp. 604–608.
6. Todhunter, I., and Leatham, J. G. *Spherical Trigonometry*, 3d ed. London: McMillan, 1911.
7. McDonald, B. H., Friedman, M., and Wexler, A., "Variational Solution of Integral Equations," *IEEE Transactions on Microwave Theory and Techniques*, March 1974, Vol. MTT-22, No. 3.
8. Vichnevetsky, R. *Computer Methods for Partial Differential Equations*, Vol. 1. Englewood Cliffs, N.J.: Prentice-Hall, Inc., 1981.
9. Zienkiewicz, O. C. *The Finite Element Method in Engineering Science*, 2d ed. London: McGraw-Hill, 1971.

9
STATIC MAGNETIC PROBLEM

9.1. INTRODUCTION

All previous chapters of this book have been devoted to developing algorithms for numerically solving field problems. The objective of this chapter and the next is to show how these algorithms have been used to solve certain practical numerical field problems. This is done to help readers apply the algorithms developed in this book to their own practical problems.

Most of the field problems and their solutions discussed here are drawn from papers that have been published in the last few years. No attempt has been made to be comprehensive, that is, to present all—or even most—of the good algorithms that appear in the recent literature. Nor are the algorithms presented here necessarily sophisticated or representative of the state-of-the-art. The reader is urged to understand the more straightforward algorithms before learning and using the more sophisticated ones.

Although these papers discuss the static magnetic field, the algorithms that they present can be applied to the static electric field as well. If we replace the magnetic field by the electric field, and the permeability by the permittivity, the formulation and the algorithm are the same.

As discussed in Chapter 3, any field problem is either an interior problem or an exterior problem. However, for certain practical reasons, it is necessary to discuss problems in a third category. These are exterior problems that are approximated by interior problems. Exterior problems that have complicated nonuniform media fall into this category. As we have seen in Chapters 6, 7, and 8, there are two facts that give this category its practical importance:

1. Complicated nonuniform media are more easily handled by the finite element method than by the integral equation method.
2. The finite element method is more easily applied to interior problems than to exterior problems.

Accordingly, Sections 9.2, 9.3, and 9.4 discuss static field problems that are respectively interior problems, exterior problems approximated by interior problems, and exterior problems.

9.2. INTERIOR STATIC FIELD PROBLEMS

Most practical static field problems are exterior problems. There are so few practical interior static field problems that very few are discussed in published papers, and none are reported here.

As shown in Chapter 6, the finite element method can be applied easily to interior static field problems. This method is straightforward even with a complicated nonuniform medium if it is linear. In such applications, the problem domain is a field problem domain and covers all points in space at which the field is to be computed.

As we saw in Section 8.2.3, the integral equation method is easily adapted to static interior problems, if the medium is linear, uniform, and isotropic over the entire problem domain. The advantage of this method is that the integral equation needs only to be solved over the source problem domain, which, in this case, is the boundary of the problem domain. For a three-dimensional problem domain, this can be a substantial advantage.

9.3. EXTERIOR STATIC PROBLEMS APPROXIMATED BY INTERIOR PROBLEMS

There are very significant practical advantages to approximating certain exterior problems by interior problems, as discussed in the introduction to this chapter. In this section, we consider the errors involved in this practice and cite certain field problems that have been solved using this practice, as presented in published papers.

In an exterior problem, the field problem domain comprises all of space. For the method discussed here, we take the following steps:

1. Arbitrarily, define a hypothetical boundary surface.
2. Over this boundary surface, place a certain boundary condition, which represents our best guess as to the field over it. This boundary condition can, of course, be Dirichlet or partly Dirichlet and partly Neumann.
3. Solve for the field inside this boundary, that is, solve the interior field problem.

If, for our boundary condition, we could use the *exact* field that exists on this boundary, then the solution to an interior problem would be identical to the solution to the exterior problem (inside the boundary). In practice, of course, we do not know the exact field on this boundary and can only approximate it. Thus, there is an error in our interior field solution that results from the error in this approximated boundary condition.

In Section 9.3.1, we examine the error in the interior solution that results from this error in the approximated boundary condition. In Section 9.3.2, we present and discuss certain exterior problems that have been approximated by interior problems, as they appear in published papers.

9.3.1. Error in Solution Caused by Error in Approximate Boundary Condition

In this section, we assume an exterior problem whose exact solution, γ_a, satisfies the equation

$$\nabla \cdot \mu \nabla \gamma_a = f \tag{9.3.1-1}$$

throughout all of space. In this equation, f is a known forcing function, that can vary with spatial coordinates. And μ is a positive scalar. (Since μ is a scalar, this problem implies an isotropic medium.) We approximate γ_a by γ_c, which is the solution to an interior problem over the arbitrarily chosen bounded domain D_i. The boundary of D_i is ∂D_i, over which we have prescribed an approximate boundary condition. (See Figure 9-1.) Again, γ_c satisfies the equation

$$\nabla \cdot \mu \nabla \gamma_c = f \tag{9.3.1-2}$$

over D_i and matches the approximate boundary conditions over ∂D_i. Notice that, over D_i the only difference between γ_c and γ_a is that γ_c equals the approximate boundary condition over ∂D_i, while γ_a equals the exact solution

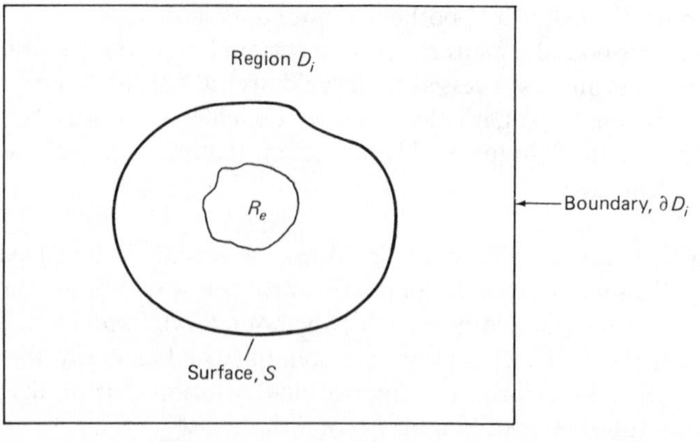

Fig. 9-1. Configuration of exterior problem approximated by interior problem.

to the exterior problem over ∂D_i. We let the error in γ_c, caused by the error in the approximate boundary condition, be γ_e, so that

$$\gamma_e = \gamma_c - \gamma_a \tag{9.3.1-3}$$

If we subtract Equation (9.3.1-2) from Equation (9.3.1-1) and use Equation (9.3.1-3), we obtain

$$\nabla \cdot \mu \nabla \gamma_e = 0 \tag{9.3.1-4}$$

Then γ_e is the solution of this equation that satisfies, over ∂D_i, a boundary condition that equals the *error* in the approximate boundary condition. And this error is the difference between the approximate boundary condition and the exact solution, γ_a, over ∂D_i.

We proceed to prove that γ_e takes its maximum and its minimum on the boundary, ∂D_i. This proof is by contradiction. Suppose that γ_e does not take its maximum on ∂D_i. Then, there must be a region (or point) R_e *inside* ∂D_i, and distinct from it, over which γ_e takes its maximum, as shown in Figure 9-1. The potentials γ_c and γ_a must both be continuous since their gradients must be finite. Since

$$\gamma_e = \gamma_c - \gamma_a$$

then γ_e must be continuous over D_i. From these facts, one can show that there must be a closed surface, S, surrounding R_e, over which

$$\mathbf{n} \cdot \nabla \gamma_e < 0$$

[as shown in Figure 9-1]. In this inequality, $\mathbf{n}$ is a unit vector directed normally outward from the surface S. Since μ must be positive, then the outflow of the vector $\mu \nabla \gamma_e$ must be negative, that is,

$$0 > \int_S \mathbf{n} \cdot \mu \nabla \gamma_e \, dS \tag{9.3.1-5}$$

However, when we use the divergence theorem in conjunction with Equation (9.3.1-4), we know

$$0 = \int_{R_e} \nabla \cdot \mu \nabla \gamma_e \, dv = \int_S \mathbf{n} \cdot \mu \nabla \gamma_e \, dS$$

and this contradicts Inequality (9.3.1-5). Therefore, the supposition that γ_e

does not take its maximum on ∂D_i is also contradicted. That is, γ_e must take its maximum on ∂D_i.

By a similar proof, we show that γ_e must also take its minimum on ∂D_i.

From these facts, we see that the maximum of $|\gamma_e|$ throughout D_i occurs on ∂D_i. In other words, if we wish to estimate the error of a solution to Equation (9.3.1-1) within a domain D_i, caused by an error in the boundary condition applied to ∂D_i, then we need only look at the error in that boundary condition.

9.3.2. Example Algorithms

Two examples of algorithms in which an exterior problem is approximated by an interior problem are presented below.

In the first of these, from a paper by Bertram and the author (Ref. 1), we compute the magnetic scalar potential in and above the pole tip of a recording head. We used a hypothetical boundary in the shape of a rectangle that encloses a pole tip and a portion of the free space immediately above the pole tip. In our application, the scalar potential is nonzero over most of the hypothetical boundary. To achieve satisfactory accuracy, we found it necessary to use care in the potential distribution that we assigned over this boundary. We plotted contours of constant scalar potential over the problem domain. We chose boundary potentials that were approximations based upon the physics of the problem. It was then apparent from our contour plots of scalar potential that these boundary potentials were sufficiently accurate.

Our second example is from a paper by Roscamp, Roberts, and Frank (Ref. 2). In it, they enclose a portion of a magnetic recording head and a portion of a magnetic medium within a rectangular hypothetical boundary. The objective of their algorithm is to compute the magnetic scalar potential of the head and the magnetization laid down in the medium. They computed the magnetic scalar potential by the finite element method. Again, they use nonzero potentials over a portion of the boundary, and, again, they had to exercise care in the choice of these nonzero boundary potentials. They did this by plotting the computed magnetic field and magnetizations, and making sure that these made sense physically.

9.4. EXTERIOR MAGNETIC STATIC PROBLEM

9.4.1. General Considerations

Several methods for solving the exterior magnetic static problem have been reported in recent published papers. No single generally accepted solution to this problem exists, and this is because of certain difficulties and complications inherent in the problem. Essentially, the difficulty arises from the need for the field to extend to infinity and, at the same time (in many problems), to accommodate a medium that can vary in a complicated way with spatial coordinates. For problems with a uniform medium, the integral equation method is an obvious choice, since its solutions extend naturally to infinity. As shown below, this method can be extended to media that have spatial variations, but only at the expense of substantial complexity. Alternatively, the finite element method accommodates medium variations readily but can be adapted to an unbounded region only with difficulty (as discussed in Chapter 7).

9.4.2. Integral Equation Algorithms

Lean and Wexler (Ref. 3) discuss the solution of the problem shown in Figure 9-2. In this configuration, all of space is divided into just two regions, R_i and R_e. Region R_i lies inside a closed surface, S, and R_e includes all of space outside R_i. The permeability is linear and isotropic throughout all of space, is uniform at the value μ_1 throughout R_i and is uniform at the value μ_2 throughout R_e.

Because of these permeability uniformities, the equivalent charge density (discussed in Chapter 2, Section 2.6) is nonzero *only* on the surface S. For this reason, this problem can be solved by the methods of Chapter 8, Section 8.2. In these methods, an integral equation is solved over a problem domain that includes *only* the surface S. There are considerable computational advantages to solving a problem over such a restricted domain, as opposed to solving for the scalar potential over a problem domain consisting of R and R_e, as with the finite element method. These advantages are in terms of a reduced number of node points, basis functions, and finite elements used in the integral equation solution.

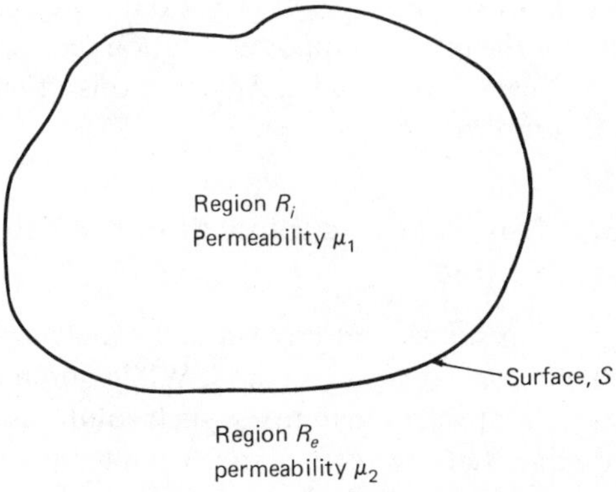

Fig. 9-2. Configuration for solution of the static magnetic problem by the boundary integral method.

Because the problem domain for the integral equation solution includes only the boundary surface, S, Lean and Wexler call this method the "boundary integral method." They show, as Section 8.2 does, that one can solve an integral equation for either the equivalent charge density over S or the scalar potential over S. From either of these solutions, one can solve for the potential either inside S, in R_i or outside S, in R_e.

In both the discussion of the boundary integral method in Section 8.2, and in Reference 3, formulations are shown for the case in which there are just two regions with the permeability uniform throughout each region. And in each case, one of these regions is bounded and the other is unbounded. However, one can see that it would be possible to extend this method to a problem that has one unbounded region and *any number* of bounded regions, again, with the permeability uniform throughout each region. Specifically, this extention can be accomplished by starting with Equations (8.2.1-2), (8.2.1-3), and (8.2.1-4) and by applying the methods used in Section 8.2.

Armstrong et al. (Ref. 4) discuss a similar formulation, in which the medium of region R_i [in Figure 9-2] is iron, which can be saturable, and the medium of region R_e is free space. In this problem, R_e is again free of real or equivalent charges. And again, there is an equivalent charge density on the surface S. However, the saturable nature of the iron in R_i makes the medium in this problem different from what it was in the Lean-Wexler problem. When saturation occurs in R_i the corresponding permeability is nonuniform. As shown in Chapter 8, Section 8.3, this nonuniformity results in a nonzero equivalent charge density over R_i. In this algorithm, then, when saturation occurs, the integral equation is solved over a problem domain that includes *both* S and R_i in Figure 9-2. They solve this integral equation for the scalar potential over this domain. Again, this solution can be used to compute the potential in R_e as well.

They also discuss the case in which saturation does not occur in R_i and the permeability is uniform in this region. They point out that in this case, the problem domain for the integral equation solution reduces to just the surface, S, as it did for the Lean-Wexler paper. Again, in this situation, the equivalent charge density is nonzero over only the surface, S.

9.4.3. General Observations Regarding the Static Magnetic Exterior Problem

The preceding section shows that any static magnetic exterior problem can be solved by the integral equation method. While nonuniform, or even nonlinear, media in such problems can make such solutions quite difficult, they are never impossible. On the other hand, Chapter 7 shows that the finite element method can also be applied to exterior problems.

Therefore, we conclude that the static magnetic exterior problem can always be solved either by the finite element method or by the integral equation method. The choice between these methods is not obvious, and is typically based on such objective considerations as the degree of nonuniformity of the medium and the preference of the user.

9.5. STATIC MAGNETIC FIELD IN A SATURABLE MEDIUM

9.5.1. General Approach

Saturable magnetic media (iron, steel, ferrite materials, etc.) are used in a number of magnetic devices (motors, generators, recording heads, etc.) in which we wish to compute magnetic fields. In computing such a field, we typically assume that the field in the medium has the following properties:

1. The medium is isotropic so that **H** and **B** are in the same direction.
2. The function **B** is a steadily increasing (nonlinear) function of **H** so that it approaches an asymtote, which is the saturation value of **B**.
3. $\nabla \cdot \mathbf{B} = 0$.
4. There is no current density in the magnetic medium, with the result that $\nabla \times \mathbf{H} = 0$. And **H** can be given in terms of the total scalar potential, ψ, by $\mathbf{H} = -\nabla\psi$.

Typically, we compute the total scalar potential ψ. One can show that ψ, computed so as to satisfy these four properties and suitably posed boundary conditions, is unique. Since the relationship between **B** and **H** is not linear, this means that the permeability is not known at the outset of the problem, at any point in the saturable medium. Typically, in fact, this permeability varies continuously from point to point throughout the saturable medium. In a saturable medium problem, then, we must compute both ψ and the permeability, μ. (Alternatively, we could think of computing both **B** and **H**.) In this way, the problem with a saturable medium is substantially more difficult than a problem with an unsaturated, or linear, medium.

Notice that μ and ψ are interrelated. For a given μ field, ψ is computed so as to satisfy assumption 3 above (as in the linear problem). But, by assumption 2, μ is a function of **H** (or $|\nabla\psi|$). Therefore, while any linear medium problem can have a *simultaneous* solution, a saturable medium problem must have an *iterative* solution. The saturable medium problem is solved by a sequence of iterations. Each iteration has two steps:

1. For a given μ field, compute a ψ field that satisfies assumptions 3 and 4.
2. For this ψ field, compute a new $\mu(\mathbf{H})$ field that satisfies assumption 2.

9.5.2. Specific Implementations

Static field computations can be implemented using either the finite element method or the integral equation method.

Bertram and the author (Ref. 1) present an algorithm based upon the numerical solution of a differential equation. While we used the finite difference method, the algorithm can be adopted to the finite element method with minor changes. This is an exterior problem that we approximated by an interior problem, as discussed in Section 9.3 above.

In Reference 5, the author adapted this algorithm to be a true solution of the exterior problem by an iterative process. This algorithm involves two iterative processes, one inside the other. The inner iteration is used to match the solution to the boundary condition at infinity. The outer iteration recomputes the permeability, μ, over the saturable medium.

Both of these algorithms were found to converge satisfactorily to a unique permeability field over the saturable field. In addition, Reference 5 proves that the iteration that matches to the boundary condition at infinity must converge.

Trowbridge (Ref. 6) presents an algorithm for computing the field in a saturable medium by the integral equation method.

REFERENCES

1. Bertram, H. Neal, and Steele, Charles W., "Pole Tip Saturation in Magnetic Recording Heads," *IEEE Transactions on Magnetics*, November 1976, Vol. MAG-12, No. G, pp. 702–706.
2. Roscamp, T. A., Roberts, G. A., and Frank, Paul D., "Optimization of Thin Film Heads for High Density Disc Recording on Particulate Media," *IEEE Transactions on Magnetics*, September 1980, Vol. MAG-16, No. 5, pp. 973–975.
3. Lean, M. H., and Wexler, A., "Accurate Field Computation with Boundary Element Method," *IEEE Transactions on Magnetics*, March 1982, Vol. MAG-18, No. 2, pp. 331–335.
4. Amstrong, A. G., Collie, C. F., Simkin, J., and Trowbridge, C. W., "The solution of 3D Magnetostatic Problems Using Scalar Potentials," Report RL-78-088, Rutherford Laboratory, Chilton, Didcot, Oxon, OX1, OQX, September 1978.
5. Steele, C. W., "Interactive Algorithm for Magnetostatic Problems with Saturable Media," *IEEE Transactions on Magnetics*, March 1982, Vol. MAG-18, No. 2, pp. 393–396.
6. Chari, M. V. K., and Silverster, P. P., Editors. *Finite Elements in Electrical and Magnetic Field Problems.* New York: John Wiley & Sons, 1980.

10
EDDY CURRENT PROBLEM

10.1. INTRODUCTION

Eddy currents are induced in any conducting medium in which the magnetic field varies with time. Thus, eddy currents occur in the stator and rotor cores of electric motors and generators, in transformers, metallic recording heads (used in magnetic recording), and the like. These eddy currents can produce undesirable effects such as power loss, heating, and—in magnetic recording—degradation in performance. We wish to compute these eddy currents in order to predict these problems in advance, before expensive prototypes are built.

In the eddy current problem, the time variation of the magnetic field induces an electric field. This electric field, in turn, causes an electric current to flow in the conducting medium. These currents, in turn, affect the magnetic field. Thus, the electric and magnetic field interact, in much the same way that the electric and magnetic fields of, say, a radio wave do. The eddy current problem is somewhat simpler than the dynamic problem, however, because displacement currents can be neglected. One can show that for all practical frequencies, and for the conductivities of all core metals, the displacement current is much smaller than the conduction current. That is, eddy current problems are *quasi-static* problems (see Chapter 2, Section 2.2).

Eddy currents cause the curl of the magnetic field to be nonzero, by Maxwell's equations (as discussed in Chapter 2). Therefore, we cannot, in general, define a scalar magnetic potential that serves as a full solution to the eddy current problem. There simply is no scalar field that represents a complete solution to this problem. Putting it another way, to solve this problem, we must solve for at least one component of a vector field.

As shown below, a certain very restricted class of problems can be solved for a single-component vector field. However, most practical problems fall outside this class.

These more complex problems can be solved only by simultaneously computing two or more quantities at each point in the problem domain. These quantities may include one, two, or three components of a vector field, and may, or may not, include a scalar potential field. At least three different methods are used to formulate these problems:

1. On physical grounds, develop an energy functional; reason that this functional has a unique minimum that is the solution. This minimum is

then found by a variational approach. This is essentially a finite element method and is reported in Reference 1.

2. On physical grounds, develop a system of differential equations that, when put in operator form, yield a symmetric operator. Using this operator, develop a functional that has a unique minimum that is the solution. This minimum is found by the variational method. Again, this is a finite element method that is discussed in Reference 2.
3. On physical grounds, develop a system of differential equations. (This can be the same system used in method 2 above.) An approximate solution to this system is formulated by a projection method. (The resulting algorithm can be the same as that obtained in method 2 above.) Again, this is a finite element method, as described in Reference 3.
4. An approximate solution to the eddy current problem can be obtained by an integral equation method, as described in Reference 4.

Of these three methods, the third, a projection method, is discussed in this chapter.

Section 10.2 presents the most commonly used basic electromagnetic formulations used for the eddy current problem. Section 10.3 discusses that simple class of problems that can be solved in terms of a single-component vector field. Section 10.4 formulates the general eddy current problem using the projection approach. Finally, Section 10.5 gives an example of the use of this formulation. That is, it uses this formulation to derive the algorithm for a specific eddy current problem.

10.2. COMMONLY USED BASIC FORMULATIONS FOR THE EDDY CURRENT PROBLEM

10.2.1. $\mathbf{A} - \phi_e$ Formulation

We start with certain equations presented in Chapter 2. From Equations (2.5.2-4), (2.3.2-6), and (2.2.2-1), we have

$$\mathbf{B} = \nabla \times \mathbf{A}$$

$$\mathbf{B} = \mu \mathbf{H}$$

and

$$\nabla \times \mathbf{H} = \mathbf{J}$$

where **B** is the magnetic flux density, **A** is the magnetic vector potential, μ is the permeability, $\mathbf{H}$ is the magnetic field, and $\mathbf{J}$ is the electric current density. These equations combine to yield

$$\nabla \times \frac{1}{\mu} \nabla \times \mathbf{A} = \mathbf{J} \tag{10.2.1-1}$$

This equation can be used at points in the problem domain of a *static* magnetic problem, at which the current density is known at the outset, that is, where this is a driving current density.

A somewhat different formulation is required for the eddy currents. An eddy current results from an electric field acting in conjunction with conductivity. The electric field, in turn, results from the time variation of the magnetic flux density. Again, referring to Equations (2.5.4-4), (2.5.4-7), and (2.5.4-9) in Chapter 2. we have

$$\mathbf{E} = \mathbf{E}_a + \mathbf{E}_d$$

$$\mathbf{E}_a = -\frac{\partial \mathbf{A}}{\partial t}$$

and

$$\mathbf{E}_d = -\nabla \phi_e$$

where ϕ_e is the reduced electric scalar potential. From these equations, we have

$$\mathbf{E} = -\frac{\partial \mathbf{A}}{\partial t} - \nabla \phi_e \tag{10.2.1-2}$$

In addition, we know that

$$\mathbf{J}_e = \sigma \mathbf{E}$$

where σ is the conductivity. From these equations, we have

$$\mathbf{J}_e = -\sigma \frac{\partial \mathbf{A}}{\partial t} - \sigma \nabla \phi_e \tag{10.2.1-3}$$

which combines with Equation (10.2.1-1) to yield

$$\nabla \times \frac{1}{\mu} \nabla \times \mathbf{A} = -\sigma \frac{\partial \mathbf{A}}{\partial t} - \sigma \nabla \phi_e \tag{10.2.1-4}$$

Equation (10.2.1-1) is then used at points in the problem domain where there are drive currents, and Equation (10.2.1-4) is used at points where there are eddy currents. For use in computer algorithms, it is necessary to combine Equations (10.2.1-1) and (10.2.1-4) into a single equation. This equation is

$$\nabla \times \frac{1}{\mu} \nabla \times \mathbf{A} + \sigma \frac{\partial \mathbf{A}}{\partial t} + \sigma \nabla \phi_e = \mathbf{J}_s \tag{10.2.1-5}$$

We assume that, at any point in the problem domain, there is either (a) a source current density, $\mathbf{J}_s$, but no eddy current density, or (b) an eddy current density but no source current density. We see that this equation satisfies both requirements, so long as we require σ to be zero at all source current points.

For a second equation, we require divergence of the eddy current density to be zero. By taking the divergence of Equation (2.2.2-1) (stated at the beginning of this section), we see that this must be true. Therefore, the divergence of Equation (10.2.1-3) is set equal to zero:

$$\nabla \cdot \left[\sigma \frac{\partial \mathbf{A}}{\partial t} + \sigma \nabla \phi_e \right] = 0 \tag{10.2.1-6}$$

The equations of this formulation are then Equations (10.2.1-5) and (10.2.1-6). For those problems in which the fields are sinusoidally time-varying, steady-state, these equations can be put into complex form:

$$\nabla \times \frac{1}{\mu} \nabla \times \mathbf{A} + j\omega\sigma\mathbf{A} + \sigma \nabla \phi_e = \mathbf{J}_s \tag{10.2.1-7}$$

and

$$\nabla \cdot [j\omega\sigma\mathbf{A} + \sigma \nabla \phi_e] = 0 \tag{10.2.1-8}$$

We require two continuity conditions of $\mathbf{A}$ and ϕ_e at interfaces. First, since the tangential component of $\mathbf{E}$ must be continuous over an interface, we have, from Equation (10.2.1-2), that the tangential component of the vector

$$\frac{\partial \mathbf{A}}{\partial t} + \nabla \phi_e$$

be continuous at the interface. Second, from Equation (10.2.1-6), we see that the normal component of the vector

$$\sigma\left[\frac{\partial \mathbf{A}}{\partial t} + \nabla\phi_e\right]$$

must be zero over the interface.

References 2 and 3 use this formulation.

10.2.2. $T - \Omega$ Formulation

This formulation takes its name from the fact that the magnetic field, **H**, is expressed as

$$\mathbf{H} = \mathbf{T} - \nabla\Omega \qquad (10.2.2\text{-}1)$$

where **T** is the *electric vector potential* and Ω is the *magnetic scalar potential.* **T** is defined in such a way that

$$\nabla \cdot \mathbf{T} = 0 \qquad (10.2.2\text{-}2)$$

As shown in Chapter 2, any vector can be decomposed into two unique vectors, one with zero divergence and one with zero curl. Therefore, **T** and $\nabla\Omega$ are unique. Then **T** is identical to $\mathbf{H}_a$, the applied magnetic field, and Ω is identical to the reduced magnetic scalar potential, ϕ_m, as defined in Chapter 2 (within a constant). The symbols **T** and Ω are used in this discussion simply because they are in common use in papers on eddy current computations.

The $\mathbf{T} - \Omega$ formulation has been discussed by Carpenter (Refs. 5, 6, and 7) and by Preston and Reece (Refs. 1 and 8).

When the Maxwell equation, Equation (2.2.2-1) (in the quasi-static regime),

$$\nabla \times \mathbf{H} = \mathbf{J}$$

is combined with Equation (10.2.2-1), we have

$$\nabla \times \mathbf{T} = \mathbf{J} \qquad (10.2.2\text{-}3)$$

Two equations are used in a $\mathbf{T} - \Omega$ solution, a vector equation and a scalar equation. The vector equation is obtained by putting the Maxwell equation [Equation (2.2.1-2)]

$$\nabla \times \mathbf{E} = -\frac{\partial \mathbf{B}}{\partial t}$$

into $\mathbf{T} - \Omega$ terms. Since

$$\mathbf{E} = \frac{\mathbf{J}}{\sigma} = \frac{1}{\sigma}\nabla \times \mathbf{T}$$

and since, with Equation (10.2.2-1),

$$\mathbf{B} = \mu\mathbf{H} = \mu(\mathbf{T} - \nabla\Omega) \tag{10.2.2-4}$$

then Equation (2.2.1-2) becomes

$$\nabla \times \left(\frac{1}{\sigma}\nabla \times \mathbf{T}\right) = -\mu\left(\frac{\partial \mathbf{T}}{\partial t} - \nabla\frac{\partial \Omega}{\partial t}\right) \tag{10.2.2-5}$$

The scalar equation is obtained by combining the Maxwell equation [Equation (2.2.1-3)]

$$\nabla \cdot \mathbf{B} = 0$$

with Equation (10.2.2-4) to yield

$$\nabla \cdot \mu(\mathbf{T} - \nabla\Omega) = 0 \tag{10.2.2-6}$$

Now consider the interface conditions imposed upon $\mathbf{T}$ and Ω. We require Ω to be continuous over the interface. Since the curl of $\mathbf{H}$ is finite [Equation (2.2.2-1)], the component of $\mathbf{H}$ that is tangential to an interface is continuous across it. From these facts and Equation (10.2.2-1), we see that the component of $\mathbf{T}$ that is tangential to an interface is continuous across it. From Equation (10.2.2-6), we see that the component of the vector $\mu(\mathbf{T} - \nabla\Omega)$ that is normal to an interface is continuous across it.

In some applications, there is no current flow on one side of an interface. On that side of the interface, then, $\mathbf{H}$ is conservative (has zero curl, in the quasi-static regime) and can be represented *entirely* as the gradient of a scalar. In these applications, it can be convenient to make $\mathbf{T}$ zero on that side of the interface. Since the component of $\mathbf{T}$ that is tangential to the interface is continuous across the interface, then this tangential component at the interface is zero on the *other* side of the interface as well. Suppose, however, that $\mathbf{T}$ has a nonzero normal to the interface on the other side. This can be accommodated by an appropriate discontinuity in the normal derivative of Ω at the interface.

Reference 5 discusses and compares both the $\mathbf{A} - \phi$ and $\mathbf{T} - \Omega$ formulations.

10.2.3. $\mathbf{R} - \psi$ Formulation

Finally, in the last few years, a third formulation of the eddy current problem has emerged (Ref. 3). This formulation is addressed specifically to the type of problem discussed at the end of the preceding section. That is, in this problem, the space on one side of the interface is current-free and **H** is conservative and represented *entirely* as the gradient of a scalar. This is the $\mathbf{R} - \psi$ formulation, which is presented in Reference 3.

With just one exception, **R** is identical to **T**, and ψ is identical to Ω. This one exception relates to the way in which we require the component of the vector $\mu(\mathbf{R} - \nabla\psi)$ normal to an interface to be continuous across it.

Specifically, we require that $\mu(\partial\psi/\partial n)$ (where this derivative is normal to the interface) be continuous across the interface. From this, we see that the component of **R** that is normal to the interface should be continuous across it. But the tangential component of **R**, like the tangential component of **T**, is continuous across the interface. Therefore, **R** is continuous across the interface. Since **R** is zero on the current-free side of the interface, then **R** must be zero at the interface and immediately on the other side of it as well.

10.3. SIMPLE TWO-DIMENSIONAL EDDY CURRENT PROBLEM

In certain eddy current problems, the fields and media are all uniform in one direction (called the *axial* direction). We then calculate the field in a plane perpendicular to the axial direction (called the *plane* of *computation*). In certain of these configurations, the electric field, **E**, the magnetic vector potential, **A**, and the current density, **J**, are all in the axial direction. The magnetic field, **H**, lies in the plane of computation.

In this configuration, since the electric field is axial and is axially uniform, its divergence is zero. Then, with Equations (2.5.4-1) and (2.5.4-10), we see that ϕ_e equals zero as well. In addition, since **A** has only the component in the axial direction, A_z, Equation (10.2.1-5) becomes

$$\nabla\cdot\left(\frac{1}{\mu}\nabla A_z\right) - \sigma\frac{\partial A_z}{\partial t} = -J_z \qquad (10.3\text{-}1)$$

This problem is solved by computing A_z from this one equation and its boundary conditions. Chapters 6 and 7 are addressed to finite element solutions of similar problems. The only difference is that the problems addressed in those chapters do not have the second term on the left side of Equation (10.3-1). However, the algorithm discussed in those chapters can be easily revised to accommodate this term.

Solutions of this problem have been reported by Hodder and Monson (Ref. 9) and the author (Ref. 10).

10.4. PROJECTION METHODS FOR GENERAL EDDY CURRENT PROBLEMS

In Chapters 5, 6, 7, 8, and 9, and in Section 10.3 of this chapter, problems are discussed for which each solution consists of a single scalar, defined over the problem domain. Eddy current problems in general (excepting the problems discussed in Section 10.3) do not fit in this category. As we saw in Section 10.2, for example, the solution to the general eddy current problem can be expressed in terms of three components of the magnetic vector potential, **A**, and the scalar electric potential, ϕ_e. That is, the eddy current problem solution could comprise, say, A_x, A_y, A_z, and ϕ_e, with all four of these quantities defined over the problem domain. If the $\mathbf{T} - \Omega$ formulation is used, then the solution might be similarly expressed in terms of **T** and Ω, again defined over the problem domain. And the same goes for the formulation of **R** and ψ.

We proceed to develop a projection method for solving these eddy current problems that is based upon the projection methods discussed in Chapter 5. In Section 10.4.1, a Hilbert space, $\mathscr{H}_e$, suitable for eddy current problems is constructed. In Section 10.4.2, a finite-dimensional subspace of $\mathscr{H}_e$ is constructed that will have as an element our numerical solution $\boldsymbol{\beta}$. Then Section 10.4.3 splits $\boldsymbol{\beta}$ into the part that is prescribed by the essential boundary conditions and the part that is computed. Finally, in Section 10.4.4 an algorithm for the solution is constructed, using the Galerkin method.

10.4.1. Hilbert Space Construction

We proceed to construct a Hilbert space, $\mathscr{H}_e$, to be used in general eddy current calculations. As shown in Chapter 5, a Hilbert space is constructed by defining (a) its elements and (b) an inner product between its elements. Each element, **w** of $\mathscr{H}_e$ consists of a three-dimensional vector, **V**, and a scalar, ρ. And **w** along with **V** and ρ are complex functions of spatial coordinates, defined over the problem domain, D. Then we express the element w as

$$\mathbf{w} = \begin{bmatrix} \mathbf{V} \\ \rho \end{bmatrix} \tag{10.4.1-1}$$

In our eddy current problems, **V** can represent the magnetic vector potential **A**, and ρ can represent the electric potential, ϕ_e. Or **V** can represent the electric vector potential **T** and ρ can represent the magnetic scalar potential, Ω, etc.

Suppose that another element of $\mathscr{H}_e$, **u**, is defined as

$$\mathbf{u} = \begin{bmatrix} \mathbf{U} \\ \sigma \end{bmatrix} \tag{10.4.1-2}$$

Then our inner product for $\mathscr{H}_e$ can be defined in terms of the inner product of **u** and **w**

$$\langle \mathbf{u}, \mathbf{w} \rangle = \int_D [\mathbf{U} \cdot \bar{\mathbf{V}} + \sigma \bar{\rho}] \, dv \tag{10.4.1-3}$$

Reference to the discussion of inner product spaces in Chapter 5 shows that this equation satisfies all of the requirements of an inner product.

10.4.2. Finite-dimensional Subspace $\mathscr{H}_n$

Chapters 4 and 5 discussed construction of a Hilbert space, $\mathscr{H}$ and a finite dimensional subspace, $\mathscr{H}_n$, for the purpose of solving field problems that had scalar solutions. That is, each element of $\mathscr{H}$ or $\mathscr{H}_n$ was a scalar field, defined over the problem domain.

However, the Hilbert space for eddy current problems, $\mathscr{H}_e$, discussed in the previous subsection is more complex. And we need a finite dimensional space $\mathscr{H}_n$, that is, a subspace of $\mathscr{H}_e$, for use in our eddy current solution. (Our numerical solution, $\boldsymbol{\beta}$ is an element of $\mathscr{H}_n$). We proceed to construct this eddy current $\mathscr{H}_n$ in much the same way that we constructed the $\mathscr{H}_n$ for scalar problems in Chapters 4 and 5.

This construction is essentially that given in Reference 3. The first two steps of this construction are those used in Chapter 4:

1. Cover the problem domain, D, with node points and finite elements, as discussed in Chapter 4. In this chapter, the total number of these node points is symbolized by M.
2. Construct functions $f_k(x, y, z)$ for $1 \leqslant k \leqslant M$. Each f_k is defined over D. That is, the domain of f_k is D. Furthermore, each f_k corresponds to one of the M node points. Finally, each f_k is, at all points in D, identical to one of the local-support pyramid-type basis functions discussed in Chapter 4 [see Figure 4-1b]. Then f_k equals unity at node point K and equals zero at all other node points.

However, the construction of the basis functions of this new $\mathscr{H}_n$ is different from that in Chapters 4 and 5. In this case, each element of $\mathscr{H}_n$ or $\mathscr{H}_e$ consists of *four* functions defined over D. (These are the scalar, ρ, and the three components of the vector, $\mathbf{V}$, that appear on the right side of Equation (10.4.1-1). Therefore, we have *four* basis functions in $\mathscr{H}_n$ corresponding to each node point. That is, there are a total of $4M$ linearly independent elements in a basis of $\mathscr{H}_n$, and the dimension of $\mathscr{H}_n$ is $4M$.

Suppose that these basis functions are $\boldsymbol{\alpha}_i$ for $1 \leqslant i \leqslant 4M$. Suppose also that $\boldsymbol{\alpha}_i$ is one of those corresponding to node point k. Then $\boldsymbol{\alpha}_i$ is given by

$$\boldsymbol{\alpha}_i = \begin{bmatrix} \mathbf{W}_i \\ \xi_i \end{bmatrix} \tag{10.4.2-1}$$

where

$$\mathbf{W}_i = \mathbf{Q}_i f_k \tag{10.4.2-2}$$

and

$$\xi_i = h_i f_k \tag{10.4.2-3}$$

In these equations, $\mathbf{Q}_i$ is a vector of three real components, and h_i is a real number.

The approximate numerical solution $\boldsymbol{\beta}$ (which is an element of $\mathscr{H}_n$) which we will compute, is then given by

$$\boldsymbol{\beta} = \sum_{i=1}^{4M} \beta_i \boldsymbol{\alpha}_i \tag{10.4.2-4}$$

where the β_i are dimensionless complex numbers. Suppose, for example, that we solve our eddy current problem by the $\mathbf{A} - \phi_e$ method. Then the approximate fields that we compute, $\mathbf{A}_a$ and ϕ_{ea}, are related to $\boldsymbol{\beta}$ by

$$\boldsymbol{\beta} = \begin{bmatrix} \mathbf{A}_a \\ \phi_{ea} \end{bmatrix} \tag{10.4.2-5}$$

We see from this and the previous equation that

$$\mathbf{A}_a = \sum_{i=1}^{4M} \beta_i \mathbf{W}_i \tag{10.4.2-6}$$

and

$$\phi_{ea} = \sum_{i=1}^{4M} \beta_i \xi_i \tag{10.4.2-7}$$

In these equations, β_i is dimensionless, $\mathbf{W}_i$ has the dimension of magnetic vector potential (as does $\mathbf{A}_a$), and ξ_i has the dimension electric scalar potential (as does ϕ_{ea}).

10.4.3. Decomposition of β in Accordance with Boundary Conditions

As in the projection method development of Chapter 5, certain terms in the summation of Equation (10.4.2-4) are treated differently from other terms, depending upon the boundary conditions that are imposed.

It is assumed here that the problem being formulated is an interior problem. Furthermore, the boundary of the problem domain, ∂D, is composed of just two parts, ∂E, over which an essential, Dirichlet, boundary condition is imposed, and ∂N, over which a natural boundary condition is imposed. That is, over ∂E, an essential Dirichlet boundary condition is imposed over *both* $\mathbf{A}$ and ϕ_e; and over ∂N, a natural boundary condition is imposed on *both* $\mathbf{A}$ and ϕ_e. These boundary conditions are depicted in Figure 10.1. The node points used in the problem fall into two categories:

1. The node points in the interior of D and on ∂N. These are called *active* node points.

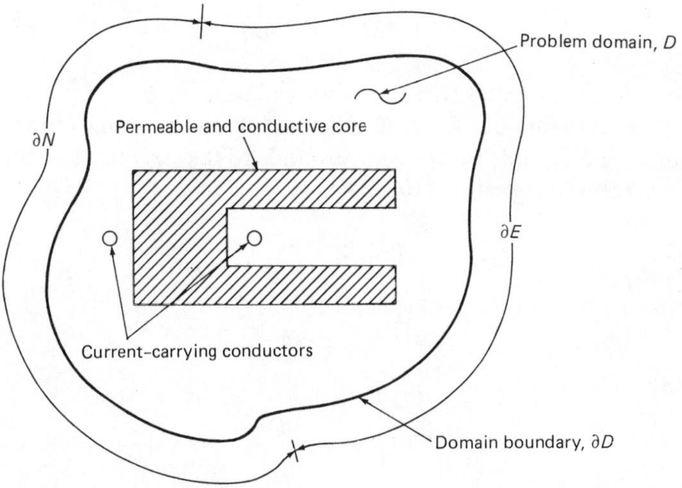

Note: In the interior eddy current problem, the current-carrying conductors, generators, and conductive and permeable cores do not *have* to be inside ∂D, as shown above.

Fig. 10-1. Interior eddy current problem.

2. The node points that fall on ∂E. These are called essential or Dirichlet node points.

The difference between these two types of node points is very significant. The fields at the active node points are all calculated in the problem, while the fields at the essential Dirichlet node points are known from the outset and need not be calculated. In the derivation given below, it is assumed that, of the M node points in the problem, M_a of these are active and M_e are essential Dirichlet, so that

$$M = M_a + M_e \tag{10.4.3-1}$$

As shown in the next section, the basis functions that correspond to active node points play a distinctly different part in the formulation from the basis functions that correspond to the essential node points. As shown in the last section, there are four basis functions for each node point. These are $4M_a$ basis functions that correspond to active node points and $4M_e$ basis functions that correspond to essential node points. For convenience, the convention is established here that the basis functions $\boldsymbol{\alpha}_i$ for

$$1 \leqslant i \leqslant 4M_a$$

correspond to active node points, and those for

$$1 + 4M_a \leqslant i \leqslant 4M$$

correspond to the essential Dirichlet node points.

Finally, we separate out $\boldsymbol{\beta}$ into $\boldsymbol{\beta}_h$, the part that corresponds to the active node points, and $\boldsymbol{\beta}_b$, the part that corresponds to the essential Dirichlet node points. Referring to Equation (10.4.2-4),

$$\boldsymbol{\beta} = \boldsymbol{\beta}_h + \boldsymbol{\beta}_b \tag{10.4.3-2}$$

$$\boldsymbol{\beta}_h = \sum_{i=1}^{4M_a} \beta_i \boldsymbol{\alpha}_i \tag{10.4.3-3}$$

$$\boldsymbol{\beta}_b = \sum_{i=1+4M_a}^{4M} \beta_i \boldsymbol{\alpha}_i \tag{10.4.3-4}$$

The significance of this split, of course, is that $\boldsymbol{\beta}_b$ is known from the outset of the problem, as soon as the boundary conditions are set and the $\boldsymbol{\alpha}_i$ are known. On the other hand, the main objective of the problem is to compute $\boldsymbol{\beta}_h$.

10.4.4. Galerkin Solution for Eddy Current Problems

We proceed to develop a formulation for eddy current problems by a projection method, specifically, the Galerkin method, described in Sections 5.4 and 5.5 of Chapter 5. Again, as in that section, the eddy current problem is first expressed by the operator equation

$$K\boldsymbol{\gamma} - \mathbf{g} = 0 \qquad (10.4.4\text{-}1)$$

In this equation $\boldsymbol{\gamma}$ and $\mathbf{g}$ are elements of the Hilbert space $\mathscr{H}_e$, constructed in Section (10.4.1). And $\boldsymbol{\gamma}$ is the exact solution to the problem, given by

$$\boldsymbol{\gamma} = \begin{bmatrix} \mathbf{A} \\ \phi_e \end{bmatrix} \qquad (10.4.4\text{-}2)$$

The operator, K, is an operator on $\mathscr{H}_e$ (that is, both D_K and R_K, the domain and range of K, are subsets of $\mathscr{H}_e$).

Notice that all elements of D_K must satisfy the boundary conditions imposed upon $\boldsymbol{\gamma}$. It is in this way that the operator, K, carries these boundary conditions.

Starting with Equation (10.4.4-1), we can write

$$\langle K\boldsymbol{\gamma}, \boldsymbol{\alpha}_i \rangle = \langle \mathbf{g}, \boldsymbol{\alpha}_i \rangle, \qquad 1 \leqslant i \leqslant 4M_a \qquad (10.4.4\text{-}3)$$

where the inner product is, of course, that defined in Section (10.4.1), and $\boldsymbol{\alpha}_i$ is a basis function of one of the *active* node points. The derivation can differ, at this point, depending upon whether an integration by parts is used. An integration by parts is assumed at this point, since an integration by parts is assumed for all finite element problems considered in this book. In general (as exemplified in the next section), the integration by parts yields:

$$\langle K\boldsymbol{\gamma}, \boldsymbol{\alpha}_i \rangle = F(\boldsymbol{\gamma}, \boldsymbol{\alpha}_i) + C(\boldsymbol{\alpha}_i), \qquad 1 \leqslant i \leqslant 4M_a \qquad (10.4.4\text{-}4)$$

where F is a bilinear form and C is a linear functional that is defined by the boundary conditions. For all of the problems considered in this book, F is symmetric. When these two equations are combined, then, we have

$$F(\boldsymbol{\gamma}, \boldsymbol{\alpha}_i) = \langle \mathbf{g}, \boldsymbol{\alpha}_i \rangle - C(\boldsymbol{\alpha}_i), \qquad 1 \leqslant i \leqslant 4M_a \qquad (10.4.4\text{-}5)$$

We require the approximate solution, $\boldsymbol{\beta}$, to satisfy this same equation, so that

$$F(\boldsymbol{\beta}, \boldsymbol{\alpha}_i) = \langle \mathbf{g}, \boldsymbol{\alpha}_i \rangle - C(\boldsymbol{\alpha}_i), \qquad 1 \leqslant i \leqslant 4M_a \qquad (10.4.4\text{-}6)$$

Equations (10.4.4-6) are then to be solved for $\boldsymbol{\beta}$. These equations give what is known as a *weak solution* (Ref. 11). From the above two equations, we have

$$F(\boldsymbol{\gamma} - \boldsymbol{\beta}, \boldsymbol{\alpha}_i) = 0, \qquad 1 \leqslant i \leqslant 4M_a$$

and from this, one can show (as was done in Chapter 5) that $\boldsymbol{\beta}$ is a *projection* of $\boldsymbol{\gamma}$ onto $\mathscr{H}_n$. From Equation (10.4.3-2),

$$\boldsymbol{\beta} = \boldsymbol{\beta}_h + \boldsymbol{\beta}_b$$

we have

$$F(\boldsymbol{\beta}_h, \boldsymbol{\alpha}_i) = \langle \mathbf{g}, \boldsymbol{\alpha}_i \rangle - C(\boldsymbol{\alpha}_i) - F(\boldsymbol{\beta}_b, \boldsymbol{\alpha}_i), \qquad 1 \leqslant i \leqslant 4M_a \qquad (10.4.4\text{-}7)$$

where only $\boldsymbol{\beta}_h$ is to be computed, and all of the terms on the right side are known at the outset of the problem. When Equations (10.4.3-4) and (10.4.3-3) are combined with this equation, we have

$$\sum_{j=1}^{4M_a} \beta_j F(\boldsymbol{\alpha}_j, \boldsymbol{\alpha}_i) = \langle \mathbf{g}, \boldsymbol{\alpha}_i \rangle - C(\boldsymbol{\alpha}_i) - \sum_{j=1+4M_a}^{4M} \beta_j F(\boldsymbol{\alpha}_j, \boldsymbol{\alpha}_i), \qquad 1 \leqslant i \leqslant 4M_a \qquad (10.4.4\text{-}8)$$

Equations (10.4.4-8) represent a linear system of order $4M_a$, which, when solved, will yield $\boldsymbol{\beta}_h$ and the solution to our problem.

For practical computations, it is convenient to put this linear system in the form of the matrix-vector equation

$$L\hat{\boldsymbol{\beta}} = \mathbf{h} \qquad (10.4.4\text{-}9)$$

where L is a square matrix of order $4M_a$, with elements given by

$$l_{ji} = F(\boldsymbol{\alpha}_j, \boldsymbol{\alpha}_i), \qquad 1 \leqslant i, j \leqslant 4M_a \qquad (10.4.4\text{-}10)$$

The right-hand vector $\mathbf{h}$ has elements

$$h_i = \langle \mathbf{g}, \boldsymbol{\alpha}_i \rangle - C(\boldsymbol{\alpha}_i) - \sum_{j=1+4M_a}^{4M} F(\boldsymbol{\alpha}_j, \boldsymbol{\alpha}_i) \qquad (10.4.4\text{-}11)$$

and the solution vector $\hat{\boldsymbol{\beta}}$ has elements β_i for $1 \leqslant i \leqslant 4M_a$ (the coefficients of the basis functions of the active node points).

10.5. EDDY CURRENT PROBLEM FORMULATION

10.5.1. Statement of the Problem

The purpose of this section is to give an example of the way in which the Hilbert space, $\mathscr{H}_a$, and the formulation developed in the previous section can be used to develop the algorithm for an eddy current problem. While the particular algorithm developed computes **A** and ϕ_e, that formulation could be used equally well to develop an algorithm that computes **T** and Ω, or **R** and ψ.

Specifically, the problem considered in this section is the same as the problem considered in Reference 3. The development given in this section essentially parallels the development given in that reference.

This is an *interior* problem, in which $\mathbf{A}_a$ and ϕ_{ea} are computed over the problem domain, D, by the finite element method, in either two or three dimensions. The domain boundary, ∂D, consists of two parts: ∂E, over which an essential, Dirichlet, boundary condition is imposed upon **A** and ϕ_e, and ∂N, over which a natural boundary condition is imposed, again, on both **A** and ϕ_e. See Figure 10.1.

Since most practical eddy current problems are exterior problems, we need to consider how this algorithm could be extended to an exterior problem. Suppose that all sources and permeable and conductive materials are confined within a sphere of finite radius, as shown in Figure 10-2. In this configuration, both ϕ_e and **A** go to zero at infinity. The algorithm can be converted from an interior problem to the exterior problem of Figure 10-2 in either of two ways:

1. One could approximate the exterior problem by putting a hypothetical boundary suitably far outside this sphere and requiring ϕ_a and $\mathbf{A}_a$ to go to zero on that boundary.
2. One could use either of the algorithms presented in Chapter 7 for extending the finite element method to exterior problems.

10.5.2. Statement as Operator Equation and Inner Product Formulation

The first step in using the formulation given in Section 10.4 is to find expressions for the operator K and the right-hand term **g** that will put this problem into the form of Equation (10.4.4-1):

$$K\gamma = \mathbf{g}$$

Using Equations (10.2.1-7) and (10.2.1-8), we have

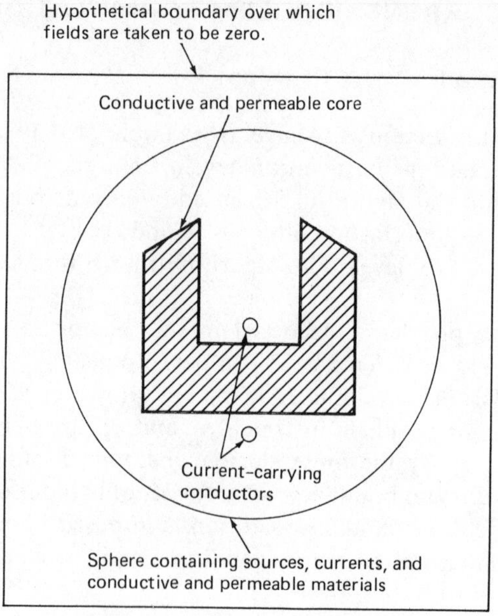

Fig. 10-2. Exterior eddy current problem.

$$K\boldsymbol{\gamma} = \begin{bmatrix} \nabla \times \dfrac{1}{\mu} \nabla \times \mathbf{A} + j\omega\sigma\mathbf{A} + \sigma\nabla\phi_e \\ -\nabla \cdot \left(\sigma\mathbf{A} + \dfrac{\sigma\nabla\phi_e}{j\omega}\right) \end{bmatrix} = \mathbf{g} = \begin{bmatrix} \mathbf{J}_s \\ 0 \end{bmatrix} \quad (10.5.2\text{-}1)$$

where (Equation 10.4.4-2)

$$\boldsymbol{\gamma} = \begin{bmatrix} \mathbf{A} \\ \phi_e \end{bmatrix}$$

The next step is to formulate the inner products as they appear in Equation (10.4.4-3). Using the definition of $\boldsymbol{\alpha}_i$ in Equation (10.4.2-1), we have

$$\langle K\boldsymbol{\gamma}, \boldsymbol{\alpha}_i \rangle = \left\langle K\boldsymbol{\gamma}, \begin{bmatrix} W_i \\ \xi_i \end{bmatrix} \right\rangle \quad (10.5.2\text{-}2)$$

and

$$\langle \mathbf{g}, \boldsymbol{\alpha}_i \rangle = \left\langle \mathbf{g}, \begin{bmatrix} W_i \\ \xi_i \end{bmatrix} \right\rangle \quad (10.5.2\text{-}3)$$

When these two equations are combined with Equation (10.5.2-1) and the inner product definition given by Equations (10.4.1-1), (10.4.2-2), and (10.4.1-3) is used, we have

$$\langle K\gamma, \boldsymbol{\alpha}_i\rangle = \int_D \left[\nabla \times \frac{1}{\mu} \nabla \times \mathbf{A} + j\omega\sigma\mathbf{A} + \sigma\nabla\phi_e\right] \cdot \mathbf{W}_i\, dv - \int_D \nabla \cdot \left(\sigma\mathbf{A} + \frac{\sigma\nabla\phi_e}{j\omega}\right) \xi_i\, dv \tag{10.5.2-4}$$

and

$$\langle \mathbf{g}, \boldsymbol{\alpha}_i\rangle = \int_D \mathbf{J}_s \cdot \mathbf{W}_i\, dv \tag{10.5.2-5}$$

10.5.3. Integration by Parts

The main function that integration by parts serves is to reduce the number of differentiations made upon the fields in Equation (10.5.2-4). Specifically, we see in this equation that in the terms

$$\nabla \times \frac{1}{\mu} \nabla \times \mathbf{A}$$

and

$$\nabla \cdot \left(\frac{\sigma\nabla\phi_e}{j\omega}\right)$$

a field has been differentiated twice. In all other terms a field is differentiated just once. We proceed to integrate these two terms by parts. By integrating these two terms by parts, we can convert Equation (10.5.2-4) into an equation in which no field is differentiated more than once. This reduction of the number of differentiations from two to one has a real advantage in the finite element method. It allows us to use shape functions that provide only continuity over the finite element boundaries. This is Lagrange interpolation, as discussed in Chapter 4. If we were to compute directly from Equation (10.5.2-4), we would need shape functions that accommodate double differentiation. These would be shape functions that provide for not only continuity, but continuity of the *normal derivative* as well, over the finite element boundaries. This is Hermite interpolation. As discussed in Chapter 4, Hermite shape functions are *not* presented in this book.

We proceed to deal with the first of the doubly differentiated terms shown above. To simplify the derivation, a new vector is defined:

$$\mathbf{H} = \frac{1}{\mu}\nabla \times \mathbf{A} \tag{10.5.3-1}$$

We use the vector identity

$$\nabla \cdot (\mathbf{W}_i \times \mathbf{H}) = \mathbf{H} \cdot (\nabla \times \mathbf{W}_i) - \mathbf{W}_i \cdot (\nabla \times \mathbf{H})$$

and integrate this equation over D to yield

$$\int_D \nabla \cdot (\mathbf{W}_i \times \mathbf{H})\, dv = \int_D \mathbf{H} \cdot (\nabla \times \mathbf{W}_i)\, dv - \int_D \mathbf{W}_i \cdot (\nabla \times \mathbf{H})\, dv$$

When the divergence theorem is applied to the left side of this equation, it becomes

$$\int_{\partial D} \mathbf{n} \cdot (\mathbf{W}_i \times \mathbf{H})\, dS = \int_D \mathbf{H} \cdot (\nabla \times \mathbf{W}_i)\, dv - \int_D \mathbf{W}_i \cdot (\nabla \times \mathbf{H})\, dv$$

where $\mathbf{n}$ is a unit normal vector. When Equation (10.5.3-1) is substituted into this equation and it is rearranged, it becomes

$$\int_D \left(\nabla \times \frac{1}{\mu}\nabla \times \mathbf{A}\right) \cdot \mathbf{W}_i\, dv = \int \left(\frac{1}{\mu}\nabla \times \mathbf{A}\right) \cdot (\nabla \times \mathbf{W}_i)\, dv$$

$$- \int_{\partial D} \mathbf{n} \cdot \mathbf{W}_i \times \left(\frac{1}{\mu}\nabla \times \mathbf{A}\right) dS \tag{10.5.3-2}$$

Next, we deal with the second of the doubly differentiated terms given above. By the vector identity

$$\nabla \cdot \xi_i(\sigma \nabla \phi_e) = \xi_i \nabla \cdot (\sigma \nabla \phi_e) + (\sigma \nabla \phi_e) \cdot \nabla \xi_i$$

we integrate this equation over D to yield

$$\int_D \nabla \cdot \xi_i(\sigma \nabla \phi_e)\, dv = \int_D \xi_i \nabla \cdot (\sigma \nabla \phi_e)\, dv + \int_D (\sigma \nabla \phi_e) \cdot \nabla \xi_i\, dv$$

When the divergence theorem is applied to the left side of this equation, it becomes (with rearrangement)

$$\int_D \xi_i \nabla \cdot (\sigma \nabla \phi_e)\, dv = \int_{\partial D} \mathbf{n} \cdot (\xi_i \sigma \nabla \phi_e)\, dS - \int_D (\sigma \nabla \phi_e) \cdot \nabla \xi_i \, dv \qquad (10.5.3\text{-}3)$$

Finally a third integration by parts is carried out for another purpose, and that is to produce a bilinear form, F, that is symmetric. While this is not essential, it is done (as in Ref. 3) because it does simplify the computation. Again, we use the vector identity

$$\nabla \cdot \xi_i(\sigma \mathbf{A}) = \xi_i \nabla \cdot (\sigma \mathbf{A}) + (\sigma \mathbf{A}) \cdot \nabla \xi_i$$

and integrating this equation over D, we have

$$\int_D \nabla \cdot \xi_i(\sigma \mathbf{A})\, dv = \int_D \xi_i \nabla \cdot (\sigma \mathbf{A})\, dv + \int_D (\sigma \mathbf{A}) \cdot \nabla \xi_i \, dv$$

When the divergence theorem is applied to the left side of this equation, and it is rearranged, it becomes

$$\int_D \xi_i \nabla \cdot (\sigma \mathbf{A})\, dv = \int_{\partial D} \mathbf{n} \cdot \xi_i \sigma \mathbf{A}\, dS - \int_D \sigma \mathbf{A} \cdot \nabla \xi_i \, dv \qquad (10.5.3\text{-}4)$$

When Equations (10.5.3-2), (10.5.3-3), and (10.5.3-4) are combined into Equation (10.5.2-4), we have

$$\langle K\boldsymbol{\gamma}, \boldsymbol{\alpha}_i \rangle = \int_D \left[\frac{1}{\mu} (\nabla \times \mathbf{A}) \cdot (\nabla \times \mathbf{W}_i) + j\omega\sigma \mathbf{A} \cdot \mathbf{W}_i \right.$$
$$\left. + \sigma(\nabla \phi_e \cdot \mathbf{W}_i + \mathbf{A} \cdot \nabla \xi_i) + \frac{\sigma}{j\omega} \nabla \phi_e \cdot \nabla \xi_i \right] dv$$
$$+ \int_{\partial D} \mathbf{n} \cdot \left[\xi_i \mathbf{A} + \frac{\sigma}{j\omega} \xi_i \nabla \phi - \mathbf{W}_i \times \left(\frac{1}{\mu} \nabla \times \mathbf{A} \right) \right] dS$$

When this equation is expressed as the sum of a bilinear form, F, and a linear functional, D, as in Equation (10.4.4-4), the result is

$$F(\boldsymbol{\gamma}, \boldsymbol{\alpha}_i) = F\left(\begin{bmatrix} \mathbf{A} \\ \phi_e \end{bmatrix}, \begin{bmatrix} \mathbf{W}_i \\ \xi_i \end{bmatrix} \right)$$
$$= \int_D \left[\frac{1}{\mu} (\nabla \times \mathbf{A}) \cdot (\nabla \times \mathbf{W}_i) + j\omega\sigma \mathbf{A} \cdot \mathbf{W}_i \right.$$
$$= \left. + \sigma(\nabla \phi_e \cdot \mathbf{W}_i + \mathbf{A} \cdot \nabla \xi_i) + \frac{\sigma}{j\omega} \nabla \phi_e \cdot \nabla \xi_i \right] dv \qquad (10.5.3\text{-}5)$$

and

$$C(\boldsymbol{\alpha}_i) = C\left(\begin{bmatrix} \mathbf{W}_i \\ \xi_i \end{bmatrix}\right)$$

$$= \int_{\partial D} \mathbf{n} \cdot \left[\xi_i \mathbf{A} + \frac{\sigma}{j\omega} \xi_i \nabla \phi_e - \mathbf{W}_i \times \left(\frac{1}{\mu} \nabla \times \mathbf{A} \right) \right] dS \qquad (10.5.3\text{-}6)$$

Notice from Equation (10.5.3-5) that the bilinear form, F, is symmetric. Furthermore, we have, using this equation that

$$F(\boldsymbol{\alpha}_j, \boldsymbol{\alpha}_i) = \int_D \left[\frac{1}{\mu} (\nabla \times \mathbf{W}_j) \cdot (\nabla \times \mathbf{W}_i) + j\omega\sigma \mathbf{W}_j \cdot \mathbf{W}_i \right.$$

$$\left. + \sigma(\mathbf{W}_i \cdot \nabla \xi_j + \mathbf{W}_j \cdot \nabla \xi_i) + \frac{\sigma}{j\omega} \nabla \xi_i \cdot \nabla \xi_j \right] dv \qquad (10.5.3\text{-}7)$$

and again, we see the symmetry in F.

In Section 10.4.4 we saw that the $\boldsymbol{\alpha}_i$ (and therefore the ξ_i and $\mathbf{W}_i$) used in these last three equations corresponds to an *active* node point, say, node point k. Then, since f_k is shaped like a pyramid, as in Figure 4-1b, it equals zero at *all other* node points. Particularly, we can say that at all essential Dirichlet node points,

$$f_k = \xi_i = \mathbf{W}_i = 0$$

Be defining ∂E to be the union of intervals *between* essential Dirichlet node points we can say that $\boldsymbol{\alpha}_i$, ξ_i, and $\mathbf{W}_i$ all equal zero over ∂E. Then the range of integration in Equation (10.5.3-6) can be reduced from ∂D to ∂N, and this equation becomes

$$C(\boldsymbol{\alpha}_i) = \int_{\partial N} \mathbf{n} \cdot \left[\xi_i \mathbf{A} + \frac{\sigma}{j\omega} \xi_i \nabla \phi_e - \mathbf{W}_i \times \left(\frac{1}{\mu} \nabla \times \mathbf{A} \right) \right] dS \qquad (10.5.3\text{-}8)$$

10.5.4. Algorithm

The algorithm is stated below, first for an interior problem and then for an exterior problem:

1. Cover the problem domain, D, with node points and finite elements, as discussed in Chapter 4. Let M be the total number of node points.

2. Choose shape functions over these finite elements. Then construct the functions $f_1, f_2, \ldots, f_M$, where each of these functions corresponds to a node point. These functions can be the same as the pyramid-type local support basis functions discussed in Chapter 4.
3. Using Equations (10.4.2-1), (10.4.2-2), and (10.4.2-3), along with the function $f_1, f_2, \ldots, f_M$ just constructed, construct the ξ_i, $\mathbf{W}_i$, and the basis functions $\boldsymbol{\alpha}_i$ for $1 \leqslant i \leqslant 4M$.
4. Using Equation (10.5.2-5), compute $\langle \mathbf{g}, \boldsymbol{\alpha}_i \rangle$ for $1 \leqslant i \leqslant 4M_a$ where M_a is the total number of active node points.
5. Using Equation (10.5.3-8), and the boundary conditions on $\mathbf{n} \cdot \mathbf{A}$, $\mathbf{n} \cdot \nabla \phi_e$, and $\mathbf{n} \cdot \mathbf{W}_i \times [(1/\mu)\nabla \times \mathbf{A}]$ over ∂N, formulate the linear functional $C(\boldsymbol{\alpha}_i)$ for $1 \leqslant i \leqslant M_a$.
6. Using Equation (10.5.3-7), compute the $F(\boldsymbol{\alpha}_j, \boldsymbol{\alpha}_i)$ for $1 \leqslant i \leqslant 4M_a$ and $1 \leqslant j \leqslant 4M$.
7. Using Equation (10.4.4-10), compute the matrix element l_{ij} for $1 \leqslant i$, $j \leqslant 4M_a$ (and therefore the matrix). Since the bilinear form F is symmetric, the matrix, L, is symmetric.
8. Using Equation (10.4.4-11), compute the vector elements h_i for $1 \leqslant i \leqslant 4M_a$ (and therefore the right-hand vector $\mathbf{h}$).
9. Using the matrix-vector Equation (10.4.4-9), compute the vector $\boldsymbol{\beta}$. This vector represents the problem solution and contains, as elements, the coefficients β_i for $1 \leqslant i \leqslant 4M_a$. This completes the solution of the interior problem.

Consider the exterior problem in which all sources and permeable and conductive materials are contained within some finite sphere, as shown in Figure 10-2. Outside this sphere, there is only free space. One way that an approximate solution to this problem can be obtained is by taking the following steps:

a. Construct a hypothetical boundary totally outside the finite sphere discussed above, as shown in Figure 10.2.
b. As a Dirichlet boundary condition, require ϕ_e and $\mathbf{A}$ to be zero over this boundary. With this boundary condition, the last two terms on the right side of Equation (10.4.4-1) disappear and this equation becomes

$$h_i = \langle \mathbf{g}, \boldsymbol{\alpha}_i \rangle, \qquad 1 \leqslant i \leqslant M_a$$

   Furthermore, it is not necessary to have *any* node points, on the hypothetical boundary, as shown in Figure 10-2. That is, $M = M_a$
c. Using these modifications, carry out the same steps 1 through 9 as used for the interior problem.

REFERENCES

1. Preston, T. W., and Reece, A. B. J., "Solution of 3-Dimensional Eddy Current Problems The T-Ω Method," *IEEE Transactions on Magnetics*, March 1982, Vol. MAG-18, No. 2, pp. 486–491.
2. Chari, M. V. K., Konrad, A., Palmo, M. A., and Angelo, J. D., "Three-Dimensional Vector Potential Analysis for Machine Field Problems," *IEEE Transactions on Magnetics*, March 1982, Vol. MAG-18, No. 2, pp. 436–446.
3. Biddlecombe, C. S., Heighway, E. A., Simkin, J., and Trowbridge, C. W., "Methods of Eddy Current Computation in Three Dimensions," *IEEE Transactions on Magnetics*, March 1982, Vol. MAG-18, No. 2, pp. 492–497.
4. McWhinter, J. H., "Numerical Solutions of Eddy Current Problems via Fredholm Equations," COMPUMAG Conference, Grenoble, France, September 1978, paper 10.3.
5. Carpenter, C. J., and Wyatt, E. A., "Efficiency of Numerical Techniques for Computing Eddy Current in Two and Three Dimensions," Proceedings COMPUMAG 76, Oxford, 1976, pp. 242–230.
6. Carpenter, C. J., "Computation of Magnetic Fields and Eddy Currents," Fifth International Conference on Magnetic Technology, Rome, April 1975, pp. 147–155.
7. Carpenter, C. J., "Comparison of Alternative Formulations of 3-Dimensional Magnetic Field and Eddy-Current Problems at Power Frequencies," *Proc IEE*, November 1977, Vol. 124, No. 11, pp. 1026–1034.
8. Preston, T. W., and Reece, A. B. J., "Finite Element Solution of Three Dimension Eddy Current Problems in Electrical Machines," Proceedings COMPUMAG, Grenoble, France, 1978, paper 7.4.
9. Hodder, W. K., and Monson, J. E., "Field Analysis for Magnetic Heads with Eddy Current or Complex Permeability," *IEEE Transactions on Magnetics*, September 1971.
10. Steele, C. W., "Convergent Algorithm for Unfounded Two-Dimensional, Linear Eddy Current Problem," COMPUMAG Conference, Grenoble, France, September 1978.
11. Milne, R. D. *Applied Functional Analysis.* Boston: Pitman Advanced Publishing Program, 1980, p. 469.

GLOSSARY

1. Greek Lower Case Letters

Letter	*Name*	*Description*
$\boldsymbol{\alpha}$	alpha	$\boldsymbol{\alpha}_1, \boldsymbol{\alpha}_2$ basis functions
$\boldsymbol{\beta}$	beta	approximate solution (linear combination of N basis functions, and therefore an element of the Hilbert space, $\mathscr{H}$)
γ	gamma	exact solution (an element of the Hilbert space, $\mathscr{H}$, and therefore a function defined over the problem domain
δ	delta	δ_{ij} Kronecker delta $\delta(i-j)$ Dirac delta function
ε	epsilon	permittivity; coordinate of a standard triangle; coordinate of a standard rectangle
ε_0		permittivity of free space
ε_r		relative permittivity
η	eta	coordinate of a standard triangle; coordinate of a standard rectangle
θ	theta	phase angle
μ	mu	permeability
μ_0		permeability of free space
$\overline{\underline{\chi_e}}$		electric susceptibility tensor
$\overline{\underline{\chi}}$		magnetic susceptibility tensor
τ	tau	higher order shape function over a triangle
ψ	psi	ψ_e total electric scalar potential; ψ_m total magnetic scalar potential
ω	omega	frequency in radians per second
σ	sigma	$\sigma(\mathbf{r}_b)$ part of mixed boundary condition; conductivity
ϕ	phi	ϕ_e electric potential, reduced; ϕ_m magnetic potential, reduced
χ	chi	χ_e electric susceptibility; χ magnetic susceptibility
ν	nu	medium parameter or medium tensor
ξ	xi	an element of S_N (not necessarily the numerical solution to the problem)
ρ	rho	volume density of electric charge; ρ_e equivalent electric charge density

2. Greek Upper Case Letters

Letter	*Name*	*Description*
Λ	lambda	triangular area coordinate
Π	pi	continued product symbol
Σ	sigma	summation
Φ	phi	magnetic flux
Ψ	psi	electric flux

3. Roman Lower Case Letters

Letter	*Description*
a	$\mathbf{a}_r$ unit vector in radial direction
b	a scalar real number
d	distance (length); $d(x, y)$ distance between points x and y
e	error (the difference between the exact solution and the approximate numerical solution
f	a function whose domain and range lie in the space of real numbers
$\mathbf{g}$	an element of the Hilbert space, (defined over the problem domain)
h	$h(r_b)$ Neumann boundary condition; h step spacing
i	integer index
j	square root of -1; integer index
k	integer index
l	length; $l_i^N(s)$ normalized Lagrange interpolating polynomial
m	$\mathbf{m}$ magnetic dipole moment; m_{ij} matrix element
n	length in the direction normal to a surface
$\mathbf{n}$	unit vector normal to surfaces
p	electric dipole moment; p_i linear interpolating function over a triangle
q	q_i linear interpolating function over a standard triangle
r	radius (distance) between two points
s	normalized distance
t	time (2.2.1)
u	an element of a linear space
v	an element of a linear space
x	a Cartesian coordinate in space
y	a Cartesian coordinate in space

z	a Cartesian coordinate in space
$\mathbf{r}_f$	radius vector from origin to field point
$\mathbf{r}_s$	radius vector from origin to source point

4. Roman Upper Case Letters

Letter	*Description*
A	area of integration magnetic vector potential
B	magnetic flux density (magnetic induction)
C	vector used in Helmholtz theorem formulation
D	displacement (electric flux density); D_k domain of the operator k
E	electric field strength; $\mathbf{E}_a$ applied electric field; $\mathbf{E}_d$ depolarizing electric field
F	force **F** vector used in Helmholtz Theorem formulation
G_2	Green's function for two-dimensional free space
G_3	Green's function for three-dimensional free space
H	magnetic field strength; $\mathbf{H}_a$ applied magnetic field; $\mathbf{H}_d$ demagnetizing magnetic field
I	electric current
J	current density; $\mathbf{J}_e$ equivalent electric current density; $\mathbf{J}_m$ equivalent magnetic current density; $\mathbf{J}_t$ total current density
K	an operator in a Hilbert space
L	triangular area coordinate; $L_i^N(x)$ Lagrange interpolating polynomial
M	number of particles;
M	magnetization;
M	number of sub-intervals into which the side of a triangle is divided
N	dimension of linear space
P	point location electric polarization
Q	electric charge
R	the residual
S	surface area; infinite-dimensional linear space
S_N	N-dimensional linear space
U	N-element vector
V	volume

W	N-element vector
X_j	cofactor of x_j in matrix M
Y_j	cofactor of y_j in matrix M
Z_j	cofactor of z_j in matrix M

APPENDIX A
DERIVATION OF THE HELMHOLTZ THEOREM

The value of the Helmholtz theorem is that it allows us to derive an integral expression for a field in terms of its divergence and its curl, which are readily obtainable from Maxwell's equations. This is done in Section 2.5.

In steps 1, 2, 3, 4, and 5 below, the Helmholtz theorem is developed in general for three dimensions. In Step 6 it is applied to certain bounded and unbounded regions. Finally, in step 7 it is adapted to two dimensions.

A.1. PRELIMINARY DERIVATION

The Helmholtz theorem is initially derived for a bounded region in three dimensions R_i, of volume V, and with a boundary surface, S, as shown in Figure A-1. As shown, there is a *source point* displaced from the origin by the radial vector $\mathbf{r}_s$. And there is a *field point*, displaced from the origin by the radial vector $\mathbf{r}_f$. The Helmholtz theorem expresses the field at the field point, $\mathbf{r}_f$ in terms of certain integrals of the field at the source points, $\mathbf{r}_s$, within the region R_i. Both $\mathbf{r}_s$ and $\mathbf{r}_f$ are defined only at points within region R_i.

To derive the Helmholtz theorem in three dimensions, the vector $\mathbf{Q}$ is defined by the volume integral

$$\mathbf{Q}(\mathbf{r}_f) = \int_V G_3(\mathbf{r}_f, \mathbf{r}_s)\mathbf{F}(\mathbf{r}_s)\, dv \tag{A-1}$$

where $G_3(\mathbf{r}_f, \mathbf{r}_s)$ is the Green's function in three-dimensional free space, as defined in Equation (2.5.1-4), $\mathbf{F}$ represents a field vector, $\mathbf{E}$, $\mathbf{H}$, $\mathbf{B}$, or $\mathbf{D}$, and V is the volume of the region R_i to which the theorem is applied. For convenience, in the derivation given below, the arguments $(\mathbf{r}_f, \mathbf{r}_s)$ of the Green's function and the argument $(\mathbf{r}_f)$ of $\mathbf{Q}$ are omitted.

If we use the vector identity

$$\nabla \times \nabla \times \mathbf{Q} = \nabla(\nabla \cdot \mathbf{Q}) - \nabla^2\mathbf{Q} \tag{A-2}$$

and we define $\mathbf{C}$ and ϕ by

$$\mathbf{C} = \nabla \times \mathbf{Q} \tag{A-3}$$

$$\phi = \nabla \cdot \mathbf{Q} \tag{A-4}$$

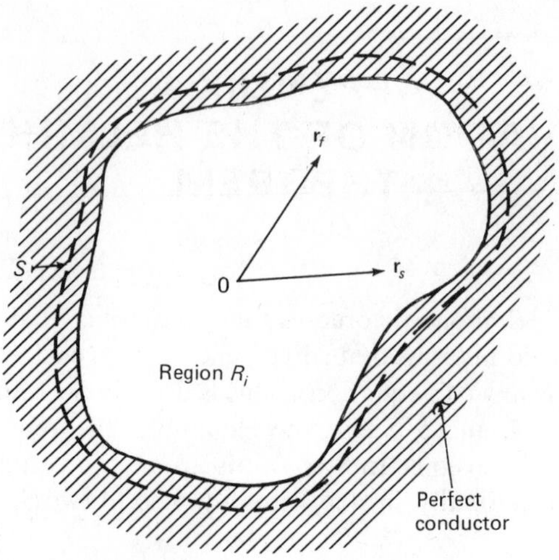

Fig. A-1. Configuration for Helmholtz theorem derivation.

then we have

$$-\nabla^2 \mathbf{Q} = \nabla \times \mathbf{C} - \nabla\phi \tag{A-5}$$

In Equations (A-2) through (A-5), the operator ∇ differentiates with respect to the field point, $\mathbf{r}_f$, while the operator ∇_s (used below) differentiates with respect to the source point, $\mathbf{r}_s$.

A.2. EVALUATION OF $\nabla^2 Q$

From Equation (A-1), since $\mathbf{F}(\mathbf{r}_s)$ is a function of the source point alone, but ∇^2 differentiates with respect to the field point,

$$\nabla^2 \mathbf{Q} = \int_V \mathbf{F}(\mathbf{r}_s) \nabla^2 G_3 \, dv \tag{A-6}$$

Since

$$\nabla^2 G_3 = 0 \qquad \text{for} \qquad \mathbf{r}_f \neq \mathbf{r}_s$$

we have with Equation (A-6),

$$\nabla^2 \mathbf{Q} = \mathbf{F}(\mathbf{r}_f) \int_V \nabla^2 G_3 \, dv$$

Since we can show by direct expansion that

$$\nabla G_3 = -\nabla_s G_3$$

and that

$$\nabla^2 G_3 = \nabla_s^2 G_3$$

we have

$$\nabla^2 \mathbf{Q} = \mathbf{F}(\mathbf{r}_f) \int_V \nabla_s^2 G_3 \, dv \tag{A-7}$$

Applying the divergence theorem to Equation (A-7), we have

$$\nabla^2 \mathbf{Q} = -\mathbf{F}(\mathbf{r}_f)$$

and with Equation (A-5),

$$\mathbf{F}(\mathbf{r}_f) = \nabla \times \mathbf{C} - \nabla \phi \tag{A-8}$$

Thus, we have decomposed **F** into a sum of irrotational and solenoidal vectors. It remains to evaluate the vector **C** and the scalar ϕ.

A.3. EVALUATION OF ϕ

Combining Equations (A-1) and (A-4),

$$\phi = \int_V \nabla \cdot [G_3 \mathbf{F}(\mathbf{r}_s)] \, dv$$

and since ∇ differentiates with respect to the field point,

$$\phi = \int_V \mathbf{F}(\mathbf{r}_s) \cdot \nabla G_3 \, dv = -\int_V \mathbf{F}(\mathbf{r}_s) \cdot \nabla_s G_3 \, dv$$

$$\phi = -\int_V \nabla_s \cdot [G_3 \mathbf{F}(\mathbf{r}_s)] \, dv + \int_V G_3 \nabla_s \cdot \mathbf{F}(\mathbf{r}_s) \, dv \tag{A-9}$$

Finally, when the divergence theorem is applied to the first term on the right side of Equation (A-9), we have

$$\phi = \int_V G_3 \nabla_s \cdot \mathbf{F}(\mathbf{r}_s)\, dv - \int_S G_3 \mathbf{F}(\mathbf{r}_s) \cdot \mathbf{n}\, dS \tag{A-10}$$

A.4. DERIVATION OF IDENTITY

In order to derive an expression for **C**, we need to derive a vector identity to be used in the next section. First, we apply the divergence theorem to the vector (**K** × **U**) where **K** is a constant vector and **U** is a variable vector:

$$\int_V \nabla_s \cdot (\mathbf{K} \times \mathbf{U})\, dv = \int_S \mathbf{n} \cdot (\mathbf{K} \times \mathbf{U})\, da \tag{A-11}$$

By manipulation of the integrals on the left and right sides of Equation (A-11), we have

$$-\mathbf{K} \cdot \int_V (\nabla_s \times \mathbf{U})\, dv = \mathbf{K} \cdot \int_S (\mathbf{U} \times \mathbf{n})\, da$$

and since this equation must hold for *any* vector **K**, we have

$$\int_V (\nabla_s \times \mathbf{U})\, dv = -\int_S (\mathbf{U} \times \mathbf{n})\, da \tag{A-12}$$

A.5. EVALUATION OF C

From Equations (A-1) and (A-3),

$$\mathbf{C} = \int_V \nabla \times [G_3 \mathbf{F}(\mathbf{r}_s)]\, dv$$

and, again, since ∇ differentiates with respect to $\mathbf{r}_f$, we have

$$\mathbf{C} = \int_V \mathbf{F}(\mathbf{r}_s) \times \nabla G_3\, dv$$

$$= -\int_V \nabla_s \times [G_3 \mathbf{F}(\mathbf{r}_s)]\, dv + \int_V G_3 \nabla_s \times \mathbf{F}(\mathbf{r}_s)\, dv \tag{A-13}$$

which shows the operations with respect to the source coordinates.

If we let

$$\mathbf{U} = G_3\mathbf{F}(\mathbf{r}_s)$$

in Equation (A-12) and combine this with Equation (A-13), we have

$$\mathbf{C} = \int_V G_3\nabla_s \times \mathbf{F}(\mathbf{r}_s)\,dv + \int_S G_3\mathbf{F}(\mathbf{r}_s) \times \mathbf{n}\,da \tag{A-14}$$

The Helmholtz theorem is stated by Equations (A-8), (A-10), and (A-14). It gives an expression for the field vector $\mathbf{F}$ in terms of volume integrals of its divergence and curl and surface integrals of its equivalent scalar and vector sources, $\mathbf{n}\cdot\mathbf{F}$ and $\mathbf{n} \times \mathbf{F}$.

A.6. APPLICATION OF THE HELMHOLTZ THEOREM TO CERTAIN BOUNDED AND UNBOUNDED REGIONS

The surface integrals in Equations (A-10) and (A-14) are usually difficult or impossible to calculate. Fortunately, there are certain regions (discussed here) for which these integrals go to zero. For this discussion, it is convenient to rewrite Equations (A-8), (A-10), and (A-14) in the following form:

$$\mathbf{F} = \nabla \times \mathbf{U} - \nabla\psi + \boldsymbol{\varepsilon}_1 - \boldsymbol{\varepsilon}_2 \tag{A-15}$$

where

$$\mathbf{U} = \int_V G_3\nabla \times \mathbf{F}\,dv \tag{A-16}$$

involves the curl of $\mathbf{F}$ and

$$\psi = \int_V G_3\nabla\cdot\mathbf{F}\,dv \tag{A-17}$$

its divergence. The remaining terms are:

$$\boldsymbol{\varepsilon}_1 = \nabla \times \int_S G_3\mathbf{F}(\mathbf{r}_s) \times \mathbf{n}\,da \tag{A-18}$$

and

$$\mathcal{E}_2 = \nabla \int_S G_3 \mathbf{F}(\mathbf{r}_s) \cdot \mathbf{n} \, da \tag{A-19}$$

First consider a bounded region R as shown in Figure A-1, where the bounding surface is S (shown as a dashed curve). The field $\mathbf{F}$ is taken to be zero over S. (This could occur, for example, if S were embedded beneath the surface of a perfect conductor, as shown in the figure). Since $\mathbf{F}$ is zero over S, we have

$$\mathcal{E}_1 = \mathcal{E}_2 = 0$$

and Equation (A-15) becomes

$$\mathbf{F} = \nabla \times \mathbf{U} - \nabla \psi \tag{A-20}$$

Now suppose that we wish to apply the Helmholtz theorem to an unbounded region, namely, all of space. To do this, we use the configuration shown in Figure A-2. In this configuration, all sources and medium discontinuities are enclosed within the sphere S_i, of radius r_i, centered at the origin. Our region R_i includes all of space inside the larger sphere, S_0, of radius r_0, centered at the origin. When the differentiations with respect to the field point in Equations (A-18) and (A-19) are carried out, we have

$$\mathcal{E}_1 = \int_{S_0} (\mathbf{F}(\mathbf{r}_s) \times \mathbf{n}) \times \nabla G_3 \, da = \int_{S_0} \frac{(\mathbf{F}(\mathbf{r}_s) \times \mathbf{n}) \times \mathbf{a}_r G_3}{R} da \tag{A-21}$$

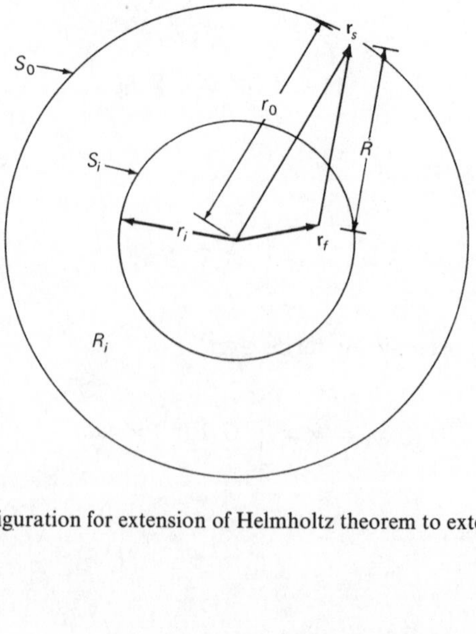

Fig. A-2. Configuration for extension of Helmholtz theorem to exterior problem.

$$\varepsilon_2 = \int_{S_O} (\mathbf{F}(\mathbf{r}_s)\cdot\mathbf{n})\nabla G_3 \, da = \int_{S_O} \frac{(\mathbf{F}(\mathbf{r}_s)\cdot\mathbf{n})\mathbf{a}_r G_3}{R} da \tag{A-22}$$

where $\mathbf{a}_r$ is a unit vector in the R direction. If we let

$$M_1 = \max_{\mathbf{r}_s \text{ in } S_O} |(\mathbf{F}(\mathbf{r}_s) \times \mathbf{n} \times \mathbf{a}_r|$$

$$M_2 = \max_{\mathbf{r}_s \text{ in } S_O} |\mathbf{F}(\mathbf{r}_s)\cdot\mathbf{n}|$$

then with (A-21) and (A-22), ε_1 and ε_2 are limited to:

$$|\varepsilon_1| \leqslant M_1 \int_{S_O} \frac{G_3}{R} da \tag{A-23}$$

$$|\varepsilon_2| \leqslant M_2 \int_{S_O} \frac{G_3}{R} da \tag{A-24}$$

Since, in the integrals in Equations (A-21) and (A-22) $\mathbf{r}_s$ is in S_O, we have

$$|\mathbf{r}_s| = r_O$$

and from Figure A-2 we have, with $\mathbf{r}_f$ inside S_i,

$$|\mathbf{r}_s - \mathbf{r}_f| = R \geqslant |\mathbf{r}_s| - |\mathbf{r}_f| = r_O - |\mathbf{r}_f| \tag{A-25}$$

Since the integrals in Equations (A-23) and (A-24) are taken over the area S_O of a sphere of radius r_O, we have with Equation (A-25) that

$$\int_{S_O} \frac{G_3}{R} da \leqslant \frac{4\pi r_O^2}{4\pi[r_O - |\mathbf{r}_f|]^2} = \frac{1}{[1-(|\mathbf{r}_f|/r_o)]^2} \tag{A-26}$$

Finally, from Equations (A-23), (A-24), and (A-25), we have that

$$|\varepsilon_1| \leqslant \frac{M_1}{[1-(|\mathbf{r}_f|/r_O)]^2} \tag{A-27}$$

$$|\varepsilon_2| \leqslant \frac{M_2}{[1-(|\mathbf{r}_f|/r_O)]^2} \tag{A-28}$$

From these equations we see that as r_O goes to infinity, since M_1 and M_2 must go to zero, so must $|\varepsilon_1|$ and $|\varepsilon_2|$. Again, in the case of an unbounded region in which all sources are confined within a bounded region, the field **F** is given by Equation (A-20), and G and ψ are given by Equations (A-16) and (A-17).

A.7. TWO-DIMENSIONAL HELMHOLTZ THEOREM

The Helmholtz theorem is frequently needed for two-dimensional problems. For the most part, these are problems with unbounded regions that may or may not have excluded from them bounded regions (over whose boundaries the field goes to zero), as in Figure A-3a. In addition, it is conceivable that one might wish to use the Helmholtz theorem in a two-dimensional bounded-region problem, as in Figure A-3b. Accordingly, a version of the Helmholtz theorem that applies to all of these regions is derived below.

Each of these regions has an axis (the z direction) along which the region, the medium, and any field sources, and the field **F** are invariant. If the region is unbounded, then we assume that there is a circular cylinder, having the z axis as its axis, outside which there are no sources or medium nonuniformities, only free space, as shown in Figure A-3a.

While in the three-dimensional unbounded-region case we could assume that there is a region outside which all sources and medium nonuniformities are zero, that is not so in the two-dimensional case, where these extend to infinity in the z direction. To handle this, we first apply the integrals in Equations (A-16) and (A-17) as if they extended only between the interval $-z_l \leq z \leq z_l$, and then we let z_l go to infinity. Equation (A-17) then becomes

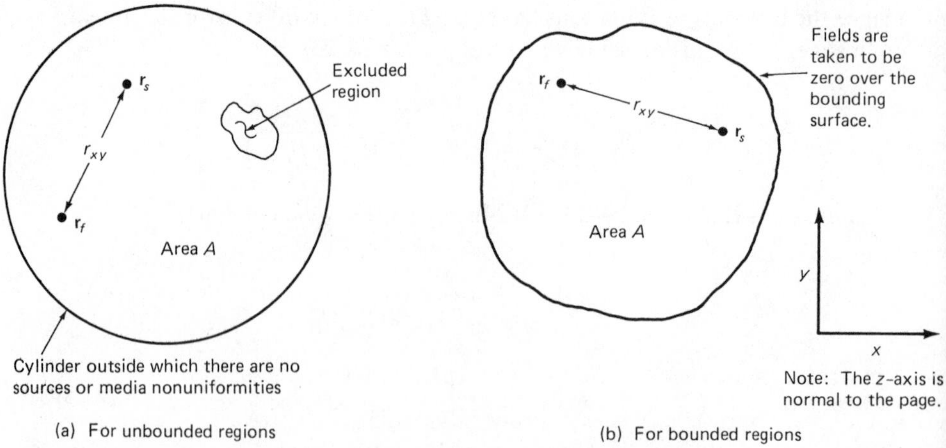

Fig. A-3. Configurations for two-dimensional Helmholtz theorem.

$$\psi = \lim_{z_l \to \infty} \int_A \int_{-z_l}^{z_l} (\nabla \cdot \mathbf{F}) G_3 \, dz \, da \tag{A-29}$$

where A is the cross-sectional area of the cylinder that contains all sources and media discontinuities. That is, $\nabla \cdot \mathbf{F}$ goes to zero outside this cylinder.

Since $\nabla \cdot \mathbf{F}$ is independent of z, we have

$$\psi = \lim_{z_l \to \infty} \int_A \nabla \cdot \mathbf{F} I(x, y) \, da \tag{A-30}$$

where

$$I(x, y) = \int_{z=-z_l}^{z_l} G_3 \, dz \tag{A-31}$$

We take the field point $\mathbf{r}_f$ to be in the $z = 0$ plane, with the result that

$$G_3 = \frac{1}{4\pi\sqrt{r_{xy}^2 + z^2}}$$

where r_{xy} equals the distance from $\mathbf{r}_f$ to the orthogonal projection of $\mathbf{r}_s$ onto the $z = 0$ plane. Since G_3 is an even function of z, we have, from Equations (A-30) and (A-31), that

$$\begin{aligned}
I(x, y) &= \frac{1}{2\pi} \int_0^{z_l} \frac{dz}{\sqrt{r_{xy}^2 + Z^2}} \\
&= \frac{1}{2\pi} [\ln(\sqrt{r_{xy}^2 + z^2} + z)]_{z=0}^{z_l} \\
&= \frac{1}{2\pi} [\ln(\sqrt{r_{xy}^2 + z_l^2} + z_l) - \ln r_{xy}] \\
&= \frac{1}{2\pi} \left[\ln 2z_l + \ln\left(\frac{1}{2}\sqrt{\frac{r_{xy}^2}{z_l^2} + 1} + \frac{1}{2} \right) - \ln r_{xy} \right]
\end{aligned} \tag{A-32}$$

Since all medium nonuniformities and sources are confined within a cylinder of cross-sectional area A, we know that

$$\int_A \nabla \cdot \mathbf{F} \, da = 0 \tag{A-33}$$

When we combine Equation (A-30) with Equation (A-32) and use Equation (A-33), only the $\ln r_{xy}$ term survives, and we obtain

$$\psi = -\frac{1}{2\pi} \int_A \nabla \cdot \mathbf{F} \ln r_{xy} \, da \tag{A-34}$$

In this equation, r_{xy} is a function of the x and y coordinates *only* of the source and field points. It might appear upon casual inspection that the integral in Equation (A-34) diverges as r_{xy} goes to zero. This is not true. One can show that this integral converges even as r_{xy} goes to infinity by using Equation (A-33) and noting that $\nabla \cdot \mathbf{F}$ is nonzero *only* over the cross-sectional area, A.

Again, since all medium nonuniformities and sources are confined within a cylinder of cross-sectional area A, we know that

$$\int_A \nabla \times \mathbf{F} \, da = 0 \tag{A-35}$$

Using this fact, we show, from Equation (A-16), in a similar way that in the two-dimensional case,

$$\mathbf{U} = -\frac{1}{2\pi} \int_A \nabla \times \mathbf{F} \ln r_{xy} \, da \tag{A-36}$$

APPENDIX B
PROPERTIES OF THE MAGNETIC VECTOR POTENTIAL, A

B.1. PROOF THAT THE DIVERGENCE OF A IS ZERO

Taking the divergence of Equation (2.5.2-3) yields

$$\nabla \cdot \mathbf{A} = \mu_0 \int_V (\mathbf{J} + \nabla_s \times \mathbf{M}) \cdot \nabla G_3 \, dv$$

By interchanging observer and source gradients via Equation (2.4.2-3),

$$\nabla \cdot \mathbf{A} = -\mu_0 \int_V (\mathbf{J} + \nabla_s \times \mathbf{M}) \cdot \nabla_s G_3 \, dv$$

and by a vector identity,

$$\nabla \cdot \mathbf{A} = -\mu_0 \int_V [\nabla_s \cdot G_3(\mathbf{J} + \nabla_s \times \mathbf{M}) - G_3 \nabla_s \cdot (\mathbf{J} + \nabla_s \times \mathbf{M})] \, dv \qquad \text{(B-1)}$$

From Equation (2.2.2-1) (expressing **J** as the curl of H), and since the divergence of the curl of any vector equals zero, we have only the first term of Equation (B-1) left,

$$\nabla \cdot \mathbf{A} = \mu_0 \int_V \nabla_s \cdot G_3(\mathbf{J} + \nabla_s \times \mathbf{M}) \, dv$$

which, with the divergence theorem, becomes

$$\nabla \cdot \mathbf{A} = \mu_0 \int_S G_3(\mathbf{J} \times \nabla_s \times \mathbf{M}) \cdot \mathbf{n} \, da \qquad \text{(B-2)}$$

In the case of a bounded region, **J** and **M** go to zero over the boundary S. In the case of an unbounded region, we can make S be a closed surface that encloses all sources and permeable, dielectric, and conductive materials. Then, again, **J** and **M** go to zero over S. In any case, from Equation (B-2),

$$\nabla \cdot \mathbf{A} = 0 \qquad \text{(B-3)}$$

B.2. ALTERNATIVE FORMULATION OF A

Applying Equation (A-12) to the vector $G_3\mathbf{M}$ yields

$$\int_V \nabla \times (G_3\mathbf{M})\,dv = -\int_S (G_3\mathbf{M}) \times \mathbf{n}\,da$$

but since $\mathbf{M}$ goes to zero on S, (provided surface S is taken sufficiently large)

$$\int_V \nabla \times (G_3\mathbf{M})\,dv = 0 \tag{B-4}$$

Using Equation (B-4) with the identity

$$G_3\nabla_s \times \mathbf{M} = -\mathbf{M} \times \nabla_s G_3 + \nabla_s \times (G_3\mathbf{M})$$

results in

$$\int_V G_3\nabla_s \times \mathbf{M}\,dv = -\int_V \mathbf{M} \times \nabla_s(G_3)\,dv \tag{B-5}$$

Finally, combining Equation (2.5.2-3) with Equation (B-5) gives us

$$\mathbf{A} = \int_V \mu_0[G_3\mathbf{J} - \mathbf{M} \times \nabla_s G_3]\,dv \tag{B-6}$$

APPENDIX C
INTEGRAL EXPRESSIONS FOR SCALAR POTENTIAL FROM GREEN'S THEOREM

C.1. GREEN'S THEOREM

Suppose we have a region R_1 bounded by the closed surface S_1 (as in Figure C-1) and that γ_1, and γ_2 are two scalar fields that are defined and twice differentiable over R_1. Applying the divergence theorem to the vectors $\gamma_1 \nabla \gamma_2$ and $\gamma_2 \nabla \gamma_1$ over R_1 gives us

$$\int_{S_1} \gamma_1 \frac{\partial \gamma_2}{\partial n} da = \int_{R_1} \nabla \cdot \gamma_1 \nabla \gamma_2 \, dv = \int_{R_1} [\gamma_1 \nabla^2 \gamma_2 + \nabla \gamma_1 \cdot \nabla \gamma_2] \, dv \quad \text{(C-1)}$$

and

$$\int_{S_1} \gamma_2 \frac{\partial \gamma_1}{\partial n} da = \int_{R_1} \nabla \cdot \gamma_2 \nabla \gamma_1 \, dv = \int_{R_1} [\gamma_2 \nabla^2 \gamma_1 + \nabla \gamma_1 \cdot \nabla \gamma_2] \, dv \quad \text{(C-2)}$$

where $\partial \gamma / \partial n$ are taken in the direction normally outward from S_1. Substracting Equation (C-2) from Equation (C-1) yields

$$\int_{S_1} \left[\gamma_1 \frac{\partial \gamma_2}{\partial n} - \gamma_2 \frac{\partial \gamma_1}{\partial n} \right] da = \int_{R_1} [\gamma_1 \nabla^2 \gamma_2 - \gamma_2 \nabla^2 \gamma_1] \, dv \quad \text{(C-3)}$$

which is *Green's second theorem.*

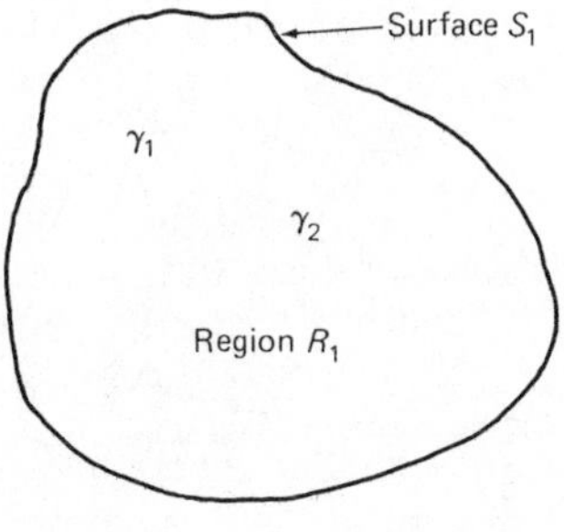

Fig. C-1. Configuration for Green's theorem derivation.

C.2. INTEGRAL EXPRESSION FOR SCALAR POTENTIAL IN A BOUNDED REGION

Suppose, now, that throughout R_1,

$$\nabla^2 \gamma_1 = 0 \tag{C-4}$$

and that

$$\gamma_2 = G_3(\mathbf{r}_f, \mathbf{r}_s) = \frac{1}{4\pi |\mathbf{r}_f - \mathbf{r}_s|} \tag{C-5}$$

As stated in Section 2.5, $G_3(\mathbf{r}_f, \mathbf{r}_s)$ is the Green's function in three-dimensional free space. Furthermore, we can show that

$$\nabla^2 G_3(\mathbf{r}_f, \mathbf{r}_s) = 0, \qquad \mathbf{r}_s \neq \mathbf{r}_f \tag{C-6}$$

Suppose now that $\mathbf{r}_s$ is *inside* R_1 or on S_1 and $\mathbf{r}_f$ is *outside* R_1 and S_1, as in Figure C-2a. Then $\mathbf{r}_s \neq \mathbf{r}_f$ and from Equations (C-5) and (C-6),

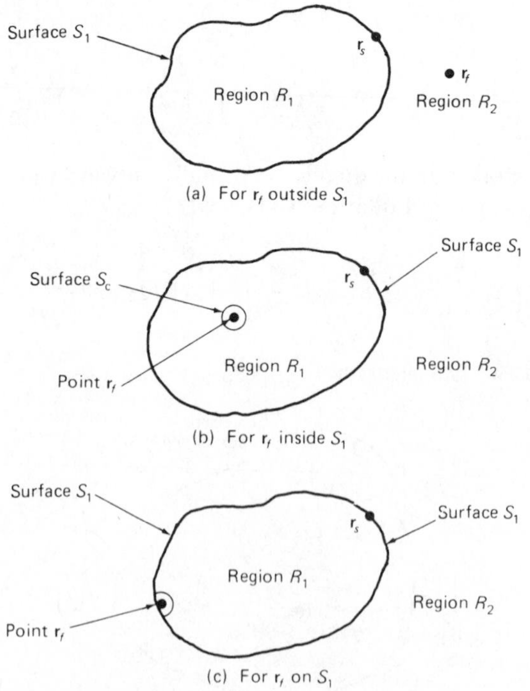

Fig. C-2. Configurations for formulating potential inside and on S_1.

$$\nabla^2 G_3 = \nabla^2 \gamma_2 = 0 \tag{C-7}$$

Assume now that $\mathbf{r}_f$ is held fixed and that the differentiations and integrations in Equation (C-3) are accomplished by variation of $\mathbf{r}_s$. Then, with Equations (C-4) and (C-7), Equation (C-3) becomes

$$\int_S \left[\gamma_1 \frac{\partial G_3}{\partial \mathrm{n}} - G_3 \frac{\partial \gamma_1}{\partial \mathrm{n}} \right] da = 0 \tag{C-8}$$

Consider now the configuration in Figure C-2b, in which $\mathbf{r}_f$ is now *inside* the bounding surface, S_1. In this configuration, there is a small sphere S_c, of radius r_c, centered on the point $\mathbf{r}_f$. In this figure, the region R_1 comprises all points *outside* S_c and *inside* or on S_1. Then R_1 now has two bounding surfaces, S_1 and S_c. Since the point $\mathbf{r}_s$ must be in R_1, then, again, $\mathbf{r}_s \neq \mathbf{r}_f$. As a result, Equation (C-8) again holds, except that the surface integration is now taken over S_1 *and* S_c, so that

$$\int_{S_1+S_c} \left[\gamma_1 \frac{\partial G_3}{\partial \mathrm{n}} - G_3 \frac{\partial \gamma_1}{\partial \mathrm{n}} \right] da = 0 \tag{C-9}$$

It remains to evaluate the surface integral over S_c. First, note that over S_c,

$$r_c = |\mathbf{r}_s - \mathbf{r}_f|$$

and G_3 and $\partial G_3/\partial \mathrm{n}$ are constant, taking the values

$$G_3 = \frac{1}{4\pi |\mathbf{r}_f - \mathbf{r}_s|} = \frac{1}{4\pi r_c} \tag{C-10}$$

and

$$\frac{\partial G_3}{\partial \mathrm{n}} = -\frac{\partial G_3}{\partial r_c} = \frac{1}{4\pi r_c^2} \tag{C-11}$$

Notice that $\mathbf{n}$ is directed normally inward, toward $\mathbf{r}_f$. Since G_3 is constant over S_c, then

$$\int_{S_c} G_3 \frac{\partial \gamma_1}{\partial \mathrm{n}} da = G_3 \int_{S_c} \frac{\partial \gamma_1}{\partial \mathrm{n}} da = G_3 \int_{V_s} \nabla^2 \gamma_1 \, dv = 0 \tag{C-12}$$

where V_s is the volume of the sphere, S_c. Then, with Equation (C-11),

$$\int_{S_c} \gamma_1 \frac{\partial G_3}{\partial n} da = \frac{1}{4\pi r_c^2} \int_{S_c} \gamma_1 \, da$$

and from this

$$\lim_{r_c \to 0} \int_{S_c} \gamma_1 \frac{\partial G_3}{\partial n} da = \gamma_1(\mathbf{r}_f) \tag{C-13}$$

When Equations (C-12) and (C-13) are combined into Equation (C-9), we have

$$\gamma_1(\mathbf{r}_f) = \int_{S_1} \left[G_3 \frac{\partial \gamma_1}{\partial n} - \gamma_1 \frac{\partial G_3}{\partial n} \right] da \tag{C-14}$$

for $\mathbf{r}_f$ totally *inside* S_1, as shown in Figure C-2b.

Finally, we have the configuration in Figure C-2c, in which $\mathbf{r}_f$ is now *on* the bounding surface, S_1. In this configuration, S_c is now a small *hemisphere*, of radius r_c centered on point $\mathbf{r}_f$. The region R_1 comprises all points *outside* S_c and on or *inside* S_1. Again, R_1 has the two bounding surfaces, S_1 and S_c and Equation (C-9) still holds. However, S_c is now a hemisphere, with a surface area of $2\pi r_c^2$. For this reason, Equation (C-13) becomes, in this application,

$$\lim_{r_c \to 0} \int_{S_c} \gamma_1 \frac{\partial G_3}{\partial n} da = \tfrac{1}{2}\gamma_1(\mathbf{r}_f) \tag{C-15}$$

And Equation (C-14) becomes

$$\tfrac{1}{2}\gamma_1(\mathbf{r}_f) = \int_{S_1} \left[G_3 \frac{\partial \gamma_1}{\partial n} - \gamma_1 \frac{\partial G_3}{\partial n} \right] da \tag{C-16}$$

Because of the similarities among Equations (C-8), (C-14), and (C-16), these equations can be conveniently combined in the form

$$\alpha_1 \gamma_1(\mathbf{r}_f) = \int_{S_1} \left[G_3 \frac{\partial \gamma_1}{\partial n} - \gamma_1 \frac{\partial G_3}{\partial n} \right] da \tag{C-17}$$

where α_1 is given by

$$\alpha_1 = \begin{cases} 0 & \mathbf{r}_f \text{ outside } S_1 \\ \frac{1}{2} & \mathbf{r}_f \text{ on } S_1 \\ 1 & \mathbf{r}_f \text{ inside } S_1 \end{cases} \tag{C-18}$$

Furthermore, it is convenient for the algorithm development that is shown in Chapter 8 to make further simplifications in the way that Equation (C-17) is expressed. Let

$$\gamma_1' = \frac{\partial \gamma_1}{\partial n} \tag{C-19}$$

and define the linear operators K and L by

$$K\gamma_1' = \int_{S_1} \gamma_1' G_3 \, da \tag{C-20}$$

$$L\gamma_1 = \int_{S_1} \gamma_1 \frac{\partial G_3}{\partial n} da \tag{C-21}$$

We call K a *single-layer* kernel operator and L a *double-layer kernel operator.* By combining Equations (C-17) and Equations (C-19) through (C-21), we have

$$\alpha_1 \gamma_1(\mathbf{r}_f) = K\gamma_1' - L\gamma_1 \tag{C-22}$$

C.3. INTEGRAL EXPRESSION FOR SCALAR POTENTIAL IN AN UNBOUNDED REGION

In Figure C-3, all points outside the surface S_1 and inside S_0 (with the possible exception of those inside S_c) constitute region R_2 as shown. Here, S_0 is a very large sphere, of radius r_0, outside S_1. We proceed to formulate γ_2 in Region R_2.

In this configuration, we assume that the only generators are at infinity. That is, all generators are outside S_0, no matter how large r_0 is. We assume that the applied static vector field produced by the generator, $\mathbf{E}_a$ or $\mathbf{H}_a$, is conservative and can therefore be represented as the gradient of a scalar potential, γ_a.

The difference between the total field, γ_2, and the applied field, γ_a, is ϕ_2, called the *scattered field*, so that

$$\phi_2 = \gamma_2 - \gamma_a \tag{C-23}$$

We assume that ϕ_2, γ_2, and γ_a all obey Laplace's equation throughout R_2. We can think of ϕ_2 as being caused by the presence of a dielectric or permeable material in R_1. That is, ϕ_2 results from polarization or magnetization in R_1 or real or equivalent charges on S_1. Then ϕ_2 is a *reduced* scalar potential [see Section (2.5)]. Furthermore, as one can show readily from the formulation given in that section, ϕ_2 must fall off at infinity as r_0^{-2}.

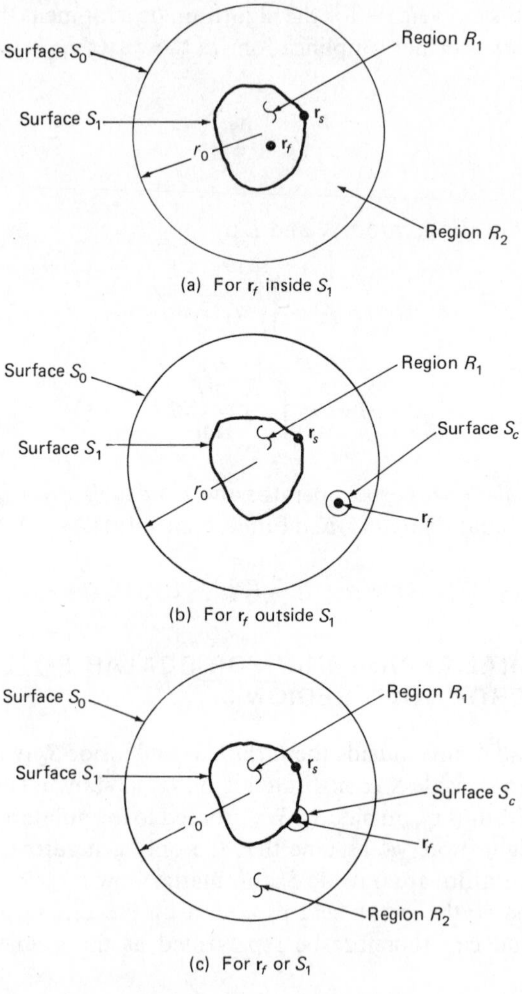

Fig. C-3. Configuration for formulating potential on and outside S_1.

Figure C-3 shows the configurations that are assumed for this derivation. Essentially, they are the reverse of the corresponding parts of Figure C-2. In this case $\mathbf{r}_s$ is always on or just *outside* S_1. Equation (C-17), when adapted to this purpose, becomes

$$\alpha_2\phi_2(\mathbf{r}_f) = \int_{S_1+S_0} \left[\phi_2 \frac{\partial G_3}{\partial n} - G_3 \frac{\partial \phi_2}{\partial n} \right] da \qquad \text{(C-24)}$$

where α_2 is given by

$$\alpha_2 = \begin{cases} 0 & \mathbf{r}_f \text{ inside } S_1 \\ \frac{1}{2} & \mathbf{r}_f \text{ on } S_1 \\ 1 & \mathbf{r}_f \text{ outside } S_1 \end{cases} \tag{C-25}$$

In Equation (C-24), the normal derivatives are taken into R_2 from S_1 and S_0. This is consistant with Equation (C-17), in which the normal derivatives are taken *out* of R_1 and *into* R_2. Since ϕ_2 falls off as r_0^{-2}, we have

$$\lim_{r_0 \to \infty} \int_{S_0} \left[G_3 \frac{\partial \phi_2}{\partial n} - \phi_2 \frac{\partial G_3}{\partial n} \right] da = 0 \tag{C-26}$$

When Equations (C-24) and (C-26) are combined and the definitions given in Equations (C-19), (C-20), and (C-21) are used, we have

$$\alpha_2 \phi_2(\mathbf{r}_f) = L\phi_2 - K\phi_2' \tag{C-27}$$

What remains is to find an expression for the total potential, γ_2, in the region R_2. Substituting Equation (C-23) into Equation (C-27) gives us

$$\alpha_2[\gamma_2(\mathbf{r}_f) - \gamma_a(\mathbf{r}_f)] = L(\gamma_2 - \gamma_a) - K(\gamma_2' - \gamma_a') \tag{C-28}$$

Now any field that is harmonic in R_1 will satisfy Equation (C-22) including γ_a, so that

$$\alpha_1 \gamma_a(\mathbf{r}_f) = K\gamma_a' - L\gamma_a \tag{C-29}$$

Subtracting Equation (C-29) from Equation (C-28), we have

$$\alpha_2 \gamma_2(\mathbf{r}_f) = L\gamma_2 - K\gamma_2' + (\alpha_1 + \alpha_2)\gamma_a(\mathbf{r}_f) \tag{C-30}$$

As shown in Figure C-3, $\mathbf{r}_f$ can be taken either inside, on, or outside S_1. Accordingly, the quantity γ_a in Equation (C-30) can be taken either inside, on, or outside S_1. In any event, however,

$$\alpha_1 + \alpha_2 = 1$$

and this equation becomes

$$\alpha_2 \gamma_2(\mathbf{r}_f) = L\gamma_2 - K\gamma_2' + \gamma_a(\mathbf{r}_f) \tag{C-31}$$

INDEX

Active node point, 76

Barycentric coordinates, 44
Basis functions, 36
  intersecting, 99
  local support of, 36
  non-intersecting, 99
  pulse, 37
  pyramid, 37
  support of, 36
Biot-Savart law, 11
Boundary conditions, 33
  Dirichlet, 33, 92, 93
  Dirichlet and Neumann boundary conditions as applied to sample linear static magnetic interior problem, 106
  essential, 74
  essential and natural boundary conditions as applied to general eddy current problem, 185
  essential, Dirichlet and natural, mixed boundary conditions in solution of linear scalar potential interior problem by the finite element method, 93
  mixed, 34, 92
  natural, 74
  Neumann, 34
  vector, 34
Bounded domain problem, 29

Collocation approach to the integral equation method, 73, 78, 149
Conduction current, 4
  density, 4
Continuity of fields at a medium discontinuity, 23
  of electric field, 23
  of magnetic field, 26
  of magnetic scalar potential, 27
  of magnetic vector potential, 27
Coulomb's law, 9
Current density, 4
  conduction, 4
  equivalent, 15

Direct sum linear space, 80
Dirichlet node point, 76
Disjoint linear spaces, 80
Displacement current, 5
Displacement density, D, 4, 7
Double-layer kernal operator, L, 217
Dynamic case for Maxwell's equations, 3

Eddy current problem, 175
  A– ϕ_e formulation, 176
  R– Ψ formulation, 181
  T– Ω formulation, 179
  two-dimensional, 181
Electric charge, 4
Electric charge density, 4
  equivalent electric charge density, 18
Electric dipole moment, 6
Electric field, 4
  applied, 18
  depolarizing, 18
  integral expression for, 17
Electric flux, 4
  density, 4
Electric scalar potential, 18
  reduced, 18
  total, 18
Electric susceptibility, 7
  tensor, 7
Electrostatic potential, 9
Equivalent configurations, 20
  for B field, 20
  for D field, 21
  for E field, 21
  for H field, 20
Equivalent surface charge density, 138
Exterior problem, 29

Field point, 9, 12, 201
Field problem, 28
 approximate, 28
 domain, 29
 real, 28
Finite element method, 73, 92, 111
 for exterior problems, 111
 for interior problems, 73, 92
Finite elements, 36
 in three-dimensional problem domains, 55
 in two-dimensional problem domains, 42

Galerkin, 60
 approach to finite element interior problem, 74, 91
 approach to integral equation problem, 78
 Bubnov-Galerkin method, 72
Gauss-Quadrature method of numerical integration, 155
 over a rectangle, 156
 over a triangle, 157
Greens function, 13
 for three-dimensional free space, 13
 for two-dimensional free space, 13
Greens theorem, 134, 213

Helmholtz theorem, 12, 134
 derivation, 201
Hermite interpolation, 41
Hilbert space, 64
 for general eddy current problems, 182
 for linear interior static potential problem, 95
 for McDonald-Wexler algorithm for linear exterior static potential problem, 113
 for sample linear interior static magnetic potential problem, 107
 for use in formulating a linear problem to be solved by either the finite element method or the integral equation method, 66

Inner product space, 63
Integral equation method, 77, 134
Integral expressions, 14
 for B field, 14
 for D field, 19
 for E field, 17
 for H field, 15
Integration by parts, 74, 94, 191
Interior problem, 29
Interpolation, 41
 Lagrange, 41
 Hermite, 41
Isoparametric shape functions in two dimensions, 51
 over a quadrangle, 53
 over a triangle, 52
Isotropic medium, 32

Lagrange, 41
 interpolating polynomials, 46
 interpolation, 41
Least squares, method of, 73
Linear medium, 32
Linear projection, 59, 77
Linear space, 35
 finite-dimensional, S_n, 35
 Hilbert space, 64
 infinite-dimensional, S, 35
 inner product, 63
 metric, 61
 normed linear, 62

Magnetic charge density, equivalent, 16
Magnetic field, 4
 applied, 16
 demagnetizing, 16
 integral expression for, 14
Magnetic flux, 4
Magnetic flux density, 4
 integral expression for, 14
Magnetic scalar potential, 16
 reduced, 16
 total, 16
Magnetic susceptibility, 9
 tensor, 8
Magnetic vector potential, quasi-static, 14
Magnetization, 8
Maxwell, J.C., 1
Maxwell's equations, 3
 dynamic case, 3
 integral forms, 4
 quasi-static case, 5
 static case, 5
Medium, 31
 isotropic, 32
 linear, 32
 uniform, 33

Metric space, 61
completeness of, 61
Minimal functional theorem, 68
Mobius coordinates, 44
Moments, methods of, 72

Newton-Coates, method of numerical integration, 155
Node point, 36
active, 76
Dirichlet, 76
Normed linear space, 62
Numerical integration, 155
Gauss-Quadrature method, 155
Newton-Coates, method, 155

Operator, 59, 64
differential, 73
domain of, 65
equation, 59, 64
integral, 78
range of, 65
Orthogonal projection, 59
Orthogonality between two elements of an inner product (or Hilbert) space, 64

Permeability, 9
relative, 9
Permittivity, 7
free space, 7
relative, 7
scalar, 7
Phasor field representations, 22
Polarization, 6
Problem domain, 30
field, 29
source, 30
Projection, 80
linear, 80
operator, 81
orthogonal, 85

Quasi-static case for Maxwell's equations, 5

Rayleigh-Ritz, 60, 68, 71

Shape function, 36
higher order, 39
higher order, over a triangular finite element, 47
in three-dimensional problem domains, 55
in two-dimensional problem domains, 42
isoparametric, in two dimensions, 51
linear, over a tetrahedron, 56
linear, over a triangle, 43
lowest order, 39
over a rectangular finite element, 49
over a rectangular parallelopiped (box), 56
Single-layer kernal operator, K, 217
Solution, 28
exact, 28
approximate numerical, 28
Source point, 9, 12, 201
Source problem domain, 30
Sparse matrix, 100
Static case for Maxwell's equations, 5
Steady-state dynamic problem, 21
Support of basis functions, 36
local, 36

Triangular area coordinates, 44
Two-dimensional problem, 31
finite elements for, 42
shape functions for, 42

Uniform medium, 33

Variational method, 70

Weighted residuals, methoa of, 73